PRECISION AGRICULTURE: SPATIAL AND TEMPORAL VARIABILITY OF ENVIRONMENTAL QUALITY

The Ciba Foundation is an international scientific and educational charity (Registered Charity No. 313574). It was established in 1947 by the Swiss chemical and pharmaceutical company of CIBA Limited — now Ciba-Geigy Limited. The Foundation operates independently in London under English trust law.

The Ciba Foundation exists to promote international cooperation in biological, medical and chemical research. It organizes about eight international multidisciplinary symposia each year on topics that seem ready for discussion by a small group of research workers. The papers and discussions are published in the Ciba Foundation symposium series. The Foundation also holds many shorter meetings (not published), organized by the Foundation itself or by outside scientific organizations. The staff always welcome suggestions for future meetings.

The Foundation's house at 41 Portland Place, London W1N 4BN, provides facilities for meetings of all kinds. Its Media Resource Service supplies information to journalists on all scientific and technological topics. The library, open five days a week to any graduate in science or medicine, also provides information on scientific meetings throughout the world and answers general enquiries on biomedical and chemical subjects. Scientists from any part of the world may stay in the house during working visits to London.

The European Environmental Research Organisation (EERO) was established in 1987 by a group of scientists from seven European countries. The members, now drawn from 14 countries, are scientists elected in recognition of their outstanding contributions to environmental research or training. EERO became fully operational in 1990 as a non-profit making, non-political foundation, with its offices at the Agricultural University of Wageningen, Generaal Foulkesweg 70, 6703 BW Wageningen, The Netherlands, EERO seeks to strengthen environmental research and training in Europe and neighbouring countries, from the Atlantic to the Urals, and from Scandinavia to the Mediterranean. It does so by promoting the most effective use of intellectual and technological resources and by supporting new and emerging scientific opportunities. EERO also provides a resource for authoritative but independent scientific assessment of environmental issues and institutions. Particular objectives are: to develop the potential of promising younger scientists by providing opportunities for collaborative research and training; to bring together participants from the relevant basic sciences to form interdisciplinary environmental research groups; to catalyse the formation and evolution of international networks to exploit emerging opportunities in environmental research; to facilitate the rapid spread in Europe of new theoretical knowledge from all parts of the world; to provide training in new developments in environmental science for those in government, industry and the research community; and to make independent scientific assessment of environmental problems and institutions, as a basis for advice to government or industry.

Ciba Foundation Symposium 210

PRECISION AGRICULTURE: SPATIAL AND TEMPORAL VARIABILITY OF ENVIRONMENTAL QUALITY

1997

JOHN WILEY & SONS

Chichester · New York · Weinheim · Brisbane · Singapore · Toronto

Copyright © Ciba Foundation 1997
Published in 1997 by John Wiley & Sons Ltd,
Baffins Lane, Chichester,
West Sussex PO19 1UD, England

National 01243 779777
International (+44) 1243 779777
e-mail (for orders and customer service enquiries): cs-books@wiley.co.uk
Visit our Home Page on http://www.wiley.co.uk
or http://www.wiley.com

All Rights Reserved. No part of this book may be reproduced, stored in a retrieval system, or transmitted, in any form or by any means, electronic, mechanical, photocopying, recording or otherwise, except under the terms of the Copyright, Designs and Patents Act 1988 or under the terms of a licence issued by the Copyright Licensing Agency, 90 Tottenham Court Road, London, UK W1P 9HE, without the permission in writing of the publisher.

Other Wiley Editorial Offices

John Wiley & Sons, Inc., 605 Third Avenue,
New York, NY 10158-0012, USA

WILEY–VCH Verlag GmbH Pappelallee 3,
D-69469 Weinheim, Germany

Jacaranda Wiley Ltd, 33 Park Road, Milton,
Queensland 4064, Australia

John Wiley & Sons (Asia) Pte Ltd, 2 Clementi Loop #02-01,
Jin Xing Distripark, Singapore 129 829

John Wiley & Sons (Canada) Ltd, 22 Worcester Road,
Rexdale, Ontario M9W 1L1, Canada

Ciba Foundation Symposium 210
viii+251 pages, 83 figures, 16 tables

Library of Congress Cataloging-in-Publication Data

Precision agriculture : spatial and temporal variability of
environmental quality.
p. cm. — (Ciba Foundation symposium ; 210)
"Editors: John V. Lake, Gregory R. Bock (organizers) and Jamie A. Goode"—Contents p.
Papers from a symposium held Jan. 21–23, 1997, in Wageningen, The Netherlands.
Includes bibliographical references and index.
ISBN 0-471-97455-2 (alk. paper)
1. Precision farming–Congresses. I. Lake, J. V. II. Bock, Gregory. III. Goode, Jamie. IV. Series.
S494.5'P73P74 1997
631–dc21 97–25663
CIP

British Library Cataloguing in Publication Data

A catalogue record for this book is available from the British Library

ISBN 0 471 97455 2

Typeset in 10/12pt Garamond by Dobbie Typesetting Limited, Tavistock, Devon.
Printed and bound in Great Britain by Biddles Ltd, Guildford and King's Lynn.
This book is printed on acid-free paper responsibly manufactured from sustainable forestation, for which at least two trees are planted for each one used for paper production.

Contents

Symposium on Precision agriculture: spatial and temporal variability of environmental quality, held in collaboration with the European Environmental Research Organisation, at the Hotel Nol in 't Bosch, Wageningen, The Netherlands on 21–23 January 1997

This symposium was based on a proposal by Alfred Stein and Johan Bouma

Editors: John V. Lake, Gregory R. Bock (Organizers) and Jamie A. Goode

Participants

V. Barnett Department of Mathematics, University of Nottingham, University Park, Nottingham NG7 2RD, UK

J. Bouma Department of Soil Science and Geology, PO Box 37, Wageningen Agricultural University, 6700 AA Wageningen, The Netherlands

I. Braud Laboratoire d'Etude des Transferts en Hydrologie et Environmement, C.N.R.S., UMR 5564, BP 53, F-38041, Grenoble 9, France

A. K. Bregt DLO Winand Staring Centre for Integrated Land, Soil and Water Research (SC-DLO), PO Box 125, 6700 AC Wageningen, The Netherlands

P. A. Burrough Netherlands Centre for Geoecology (ICG), Department of Physical Geography, Faculty of Geographical Sciences, PO Box 80115, 3508 TC Utrecht, The Netherlands

D. Goense Department of Agrotechnology and Physics, Wageningen Agricultural University, Agrotechnion Bomenweg 4, NL-6703 HD Wageningen, The Netherlands

J. J. Gómez-Hernández Department of Hydraulics and Environmental Engineering, Universidad Politécnica de Valencia, 46071 Valencia, Spain

P. M. Groffman Institute of Ecosystem Studies, Box AB, Millbrook, NY 12545, USA

G. Hudson Macaulay Land Use Research Institute, Craigiebuckler, Aberdeen, AB15 8QH, UK

M. J. Kropff Department of Theoretical Production Ecology, Wageningen Agricultural University, PO Box 430, 6700 AK Wageningen, The Netherlands

A. B. McBratney Department of Agricultural Chemistry and Soil Science, Australian Centre for Precision Agriculture, University of Sydney, Sydney, NSW 2006, Australia

J. L. Meshalkina Faculty of Soil Science, Moscow State University, Moscow 119899, Russia

P. Monestiez Unité de Biométrie, INRA, Domaine Saint Paul, Site Agroparc, 84914 Avignon, Cedex 9, France

D. J. Mulla Department of Soil, Water and Climate, Borlaug Hall, University of Minnesota, 1991 Upper Buford Circle, St. Paul, MN 55108–6028, USA

R. Rabbinge *(Chairman)* Scientific Council for Government Policy, 2 Plein 1813, PO Box 20004, 2500 EA The Hague, The Netherlands

D. Rasch Department of Mathematics, Agricultural University, Dreijenlaan 4, 6703 HA Wageningen, The Netherlands

P. Robert Department of Soil, Water and Climate, Borlaug Hall, University of Minnesota, 1991 Upper Buford Circle, St. Paul, MN 55108–6028, USA

S. E. Simmelsgaard Research Centre Foulum, PO Box 23, DK-8830 Tjele, Denmark

J. V. Stafford Silsoe Research Institute, Wrest Park, Silsoe, Bedford MK45 4HS, UK

A. Stein Department of Soil Science and Geology, PO Box 37, Wageningen Agricultural University, 6700 AA Wageningen, The Netherlands

H. Su *(Ciba Foundation Bursar)* Crop and Disease Management, IACR Rothamsted, Harpenden, Hertfordshire AL5 2JQ, UK

S. K. Thompson Department of Statistics, Pennsylvania State University, 326 Thomas Building, University Park, PA 16802–2111, USA

M. van Meirvenne Universiteit van Gent, Department of Soil Management, Coupure Links 653, B-9000 Gent, Belgium

M. Voltz Laboratoire de Science du Sol, INRA, 2 Place Viala, 34060 Montpellier, Cedex 1, France

R. Webster Department of Statistics, IACR Rothamsted, Harpenden, Hertfordshire AL5 2JQ, UK

Chairman's introduction

Rudy Rabbinge

Scientific Council for Government Policy, 2 Plein 1813, P.O. Box 20004, 2500 EA, The Hague, The Netherlands

To put this symposium on precision agriculture in its wider context, a 'birds-eye' view of the changes that have taken place in agriculture over the last century may help. Agricultural development in the twentieth century can be characterized in terms of five major trends:

(1) the increase in land and labour productivity;
(2) the use of external inputs;
(3) an increase in efficiency and efficacy of external inputs;
(4) an awareness of (and attempts to mitigate) the negative side-effects of agriculture; and
(5) the stimulation of uniformity in agricultural production areas.

One of the most significant developments has been the increase in land productivity. This is demonstrated in Fig. 1, which shows the relationship between land productivity and rice yields in Indonesia. In the 1950s rice yield increases were in the order of magnitude of 2.5 kg/ha per year, but these increased to 130 kg/ha per year by the middle of the 1960s. This dramatic rise in the rate of yield increases is known as the 'green revolution', and although this occurred in the 1960s in Asia, it had already taken place at the beginning of the century in parts of the industrialized world. A similar discontinuity in productivity rise took place just after World War II in the USA and Europe. The reason why this sort of discontinuity in production occurs is an important question. The clearest answer is because of the synergism between introduction of artificial fertilizers, mechanization, better crop varieties and the application of new agricultural knowledge.

Average wheat yields represent a good example of the nature of the changes that have occurred in agriculture. At the beginning of this century in Western Europe, wheat yields were in the order of 1500 kg/ha per year. Nowadays this has risen to 7000–9000 kg/ha per year. Last year (1996) in The Netherlands the average for the whole country was 9000 kg/ha per year. This sixfold increase occurred in less than 100 years, whereas it took many centuries for wheat yields to increase from 800 kg/ha per year to the level of 1500 kg/ha per year they had reached 100 years ago.

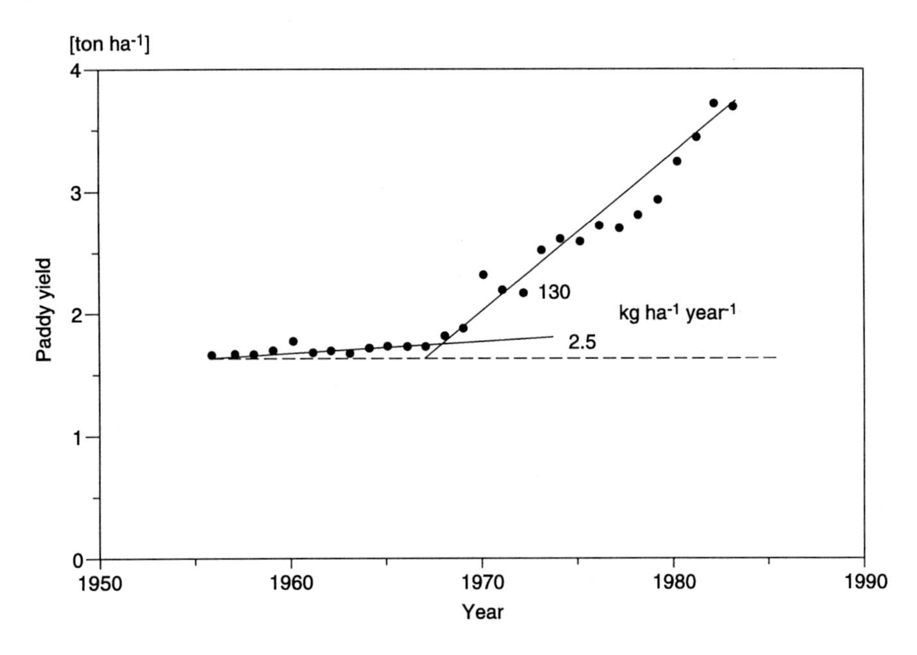

FIG. 1. Rate of increase in paddy yields in Indonesia, 1955–1980.

Even more impressive is the rise in labour productivity. At the beginning of the century a hectare of wheat required 370–400 h of labour; nowadays a similar size field needs 12–15 h labour in Western Europe and only 6 h in Australia! So, in just 100 years, labour productivity has increased 200-fold.

This increased productivity mainly has to do with applied agricultural knowledge and the use of external inputs. Primary production is influenced considerably by industrial products, such as artificial fertilizers, mechanization, water management and irrigation. It is the inter-dependence between the agricultural and industrial sectors that has helped to create this considerable increase in productivity.

The third typical characteristic is the increase in efficiency and efficacy of external inputs. This is somewhat counterintuitive. It is generally assumed that with increasing inputs the outputs increase and then plateau, as a result of diminishing returns. This holds for an individual input factor when the others are constant, but not when the inputs increase together in an orchestrated manner. Then the efficacy and efficiency of these external inputs may increase, as has been demonstrated during the last century, so instead of diminishing returns the efficiency increased. This does not continue in an unlimited way, but as long as the actual yields are still far below the potential these responses are still valid. It requires much knowledge and expertise.

Fourthly, there has been a growing awareness of (and attempts to mitigate) the negative side-effects of agriculture, caused by the over-use of external inputs and the emission of nitrates and pesticides. Broadening the objectives of agricultural production has been widely adopted. That requires again an upgrading and updating of the agricultural technologies and land use.

Finally, the stimulation of uniformity is the fifth of these characteristic trends in agriculture during this century. Through the elimination of heterogeneity at field, farm and regional level the application of simple rules and universal inputs became easy. This, however, neglects the possibility of making use of this heterogeneity.

The traditional approach of agricultural science has also changed during the last century. Previously it was based completely on 'black box' approaches at field level. It was empirical and non-explanatory, and in many cases it was and is phenomenological. This was the nature of agricultural science until the 1950s. Since then the level of explanation has increased considerably; we still have room for increasing this aspect further. In my opinion, three major innovators in agricultural science during the last century or so have contributed considerably to this development: Liebig, by introducing plant nutrition and providing the scientific basis for plant growth (plants don't eat humus but take up nitrate and other ions); Mendel, who created the science of genetics; and De Wit, who introduced systems approaches.

If we consider agriculture today we can see old concepts alongside new and emerging perspectives (Table 1). One of the old concepts is white-peg agronomy—lots of plots with many repetitions, in order to eliminate statistical variance to get a better response in terms of the relationship between the inputs and outputs. In my opinion, this outdated paradigm should be replaced by production ecological approaches in which knowledge on the biological processes involved based on insight from the physical, biological and chemical sciences is integrated through systems approaches. Thus this new concept involves the combination of experiment, simulation and understanding.

Another old concept is the production function that expresses the relationship between inputs and outputs. Agriculture very often involves the combination of

TABLE 1 Old concepts and new perspectives in agriculture

Old concept	*New perspective*	*Requirement*
White-peg agronomy	Production ecology	Training in systems approaches; simulation and exploratory studies
Production functions	Target-oriented input levels	Development of technical coefficients for production techniques
Heterogeneity as a liability	Heterogeneity as an asset	Fine-tuning of measures to specific possibilities using geographical information systems

inputs, such as nitrate, phosphate and water. The production function concept may be replaced by the target-oriented approach in which a particular production level is achieved in a specific situation by the optimal combination of various external inputs, resulting in increased yield. In this process, you define your target and see what input combination is optimal in order to reach that target. However, this requires much more knowledge and understanding.

Finally, the heterogeneity that occurs both within fields and at the regional level has often been considered to be a liability in agriculture. However, heterogeneity can be an asset if its nature is understood, because this knowledge can be put to use in tailoring agronomic measures to take account of the specific needs within a field or a farm thus facilitating more efficient use of inputs and productivity increases.

It is this latter concept that we will be concerned with during this meeting and, specifically, we will try to answer the question: how can we make use of this heterogeneity in such a way that we can further increase productivity and efficiency and efficacy of agricultural inputs? Of course, there are limits to agricultural production, but the majority of agriculture is performing at an efficiency far below potential. By making use of precision agricultural techniques we will likely be able to create more productive agricultural systems that are efficient, effective, biologically sound and which can thus contribute to the development of sustainable agricultural systems. I hope that our discussions in this meeting will contribute scientifically to the development of sustainable land use and optimal production technologies.

Precision agriculture: introduction to the spatial and temporal variability of environmental quality

Johan Bouma

Department of Soil Science and Geology, PO Box 37, Wageningen Agricultural University, 6700 AA Wageningen, The Netherlands

Abstract. Precision agriculture aims at adjusting and fine-tuning land and crop management to the needs of plants within heterogeneous fields. Production aspects have to be balanced against environmental threshold values and modern information technology has made it possible to devise operational field systems. A reactive approach is described, using yield maps and sensors. A proactive approach uses simulation modelling of plant growth and solute fluxes to predict optimal timing of management practices. Precision agriculture, combining both approaches, is seen as making a major contribution towards the development of sustainable agricultural production systems.

1997 Precision agriculture: spatial and temporal variability of environmental quality. Wiley, Chichester (Ciba Foundation Symposium 210) p 5–17

The farmer's challenge

Fields are the management units a farmer has to deal with when managing his land. Management decisions are based on either the measured or perceived conditions of the field for soils and crops at any given time, now and in the near future. Management decisions to be made by the farmer have different dimensions. First a *strategic* one: he has to make major decisions about the future thrust of his farm: for example, which crop rotation to follow, how large his herd of cows will be and which type of farm management he will pursue. Will it follow the principles of organic, biological, ecological or traditional farming? Will it be labour and capital intensive or based on part-time labour elsewhere, etc? Strategic decisions reflect his vision on the sustainability of his farming system in terms of yields, profitability, risk assessment, environmental quality and social acceptability. Then, decisions have a *tactical* dimension: for example, the choice of crop varieties or level of chemical fertilization. Finally, there are *operational* decisions: when and how much to fertilize, when to apply biocides, tillage, etc. Precision agriculture deals with all types of decisions, but its major emphasis is on operational decisions: timing and type of tillage, seeding and planting, fertilization, pest control and timing of harvest.

For chemical or organic fertilization, specific protocols have been worked out in most countries. A mixed sample from 30 surface locations evenly spread over the field is chemically analysed to arrive at a recommended single fertilization rate for the entire field, using data from standard tables which are based on dose–effect trials. Spatial variability always occurs within fields, implying occurrence of local over- and under-fertilization when the single application rate, based on the mixed sample, is applied. This is inefficient, because over-fertilization may lead to leaching of chemical fertilizer to the groundwater which is environmentally unfavourable. Decisions on tillage, seeding and planting are made by the farmer following his expert judgement rather than well established protocols: holding back sowing or planting the entire field because part of the field is too wet implies a delay for planting or sowing of parts of the field that may already have an optimal water content. Any delay usually means a shortening of the growing season and lower yields, so a trade-off decision based on economics has to be made: either go in too early in some parts of the field — with adverse effects — or too late in other parts. An even more complicated dilemma applies to tillage, which can occur in spring or autumn. Tillage under wet conditions can cause deterioration of the soil structure. Waiting in spring until the entire field is dry enough may, once again, mean an unfavourable delay in sowing or planting. Waiting in the autumn is bound to be unfavourable because the soil is likely to get wetter as time goes by. In terms of tillage, however, the farmer has — in contrast to sowing and planting — some technical alternatives, such as minimum, zero and rotation tillage.

Once a crop has been established, crop development may be hampered by many factors, such as water or nutrient shortages and the occurrence of pests and diseases. On a more mundane level, rabbits or birds may eat part of the crop or tourists may look for recreation in a wheat field. The farmer has to take corrective measures, which, when feasible, should be desirable and attractive from an economic point of view. Again, the farmer has to make many trade-offs in his decision-making process. Here also, certain protocols have been established, e.g. for spraying with biocides at pre-set times or for split applications of fertilizer. Also, sprinkling irrigation can be used to combat water shortages. In all cases, current management practices imply that measures are applied uniformly over the entire field.

The management decisions mentioned so far are made more complicated because of unknown weather conditions for the days and weeks ahead: a single shower or a dry, windy period may drastically and rapidly change conditions, further frustrating the farmer's decision-making process. This sketch of the farmer's dilemma is universal: it applies to the high-tech developed world as well as to the Third World where technical options for management are often limited but where strategic choices still have to be made by the farmer (e.g. Bouma et al 1995, 1997, Brouwer & Bouma 1997).

A major challenge for science is to present methods that can characterize variability in space and time in such a manner that farmers can use the information to improve their management. This, in fact, is the main challenge of this symposium.

Agricultural production and environmental guidelines

The examples given above are meant to illustrate that proper land management leading to sustainable agricultural production systems, which are in balance with nature and the environment, is being guided by the farmer's assessment of variation in space and time of soil and crop conditions. In the past, most emphasis has been placed on production of agricultural crops. Increasingly, however, attention is being paid to environmental side-effects of agricultural production and to product quality. A high product quality may lead to higher prices and improved profitability. Proper timing of nitrogen applications during the growing season may, for example, result in higher sugar contents of sugar beet, better malt quality of barley and a higher starch content of potatoes.

Leaching of agrochemicals to ground and surface water should be limited to quantities that do not exceed certain quality standards.

Soil quality, in this context, is defined as 'the capacity of a specific kind of soil to: (i) function within natural or managed agrosystem boundaries; (ii) sustain plant and animal production; (iii) maintain or enhance water quality; and (iv) support human health and habitation' (SSSA 1995). Soil quality is judged by threshold values for environmental indicators such as the nitrate and biocide content of groundwater and the content of heavy metals in soil. We must always look at quantity and quality of agricultural production and to environmental quality, as they are important elements of sustainable agricultural production systems.

When we fine-tune the application of agrochemicals to the needs of the growing plant, we tend to improve the efficiency of these compounds as losses are avoided. Fine-tuning, in terms of time and space, of not only the application of agrochemicals but also of all other crop management measures is the core activity of precision agriculture. We must consider the entire production system with all its interrelated processes.

The crucial role of new technology

Farmers have long recognized the importance of precision agriculture or 'smart farming' as it is sometimes called. However, the practical implementation of precision agriculture has not occurred until recently because of technical limitations: there were no affordable ways of applying management in a site-specific manner, nor were techniques well developed for defining soil quality in space and time. Now we see two rather spectacular changes:

(1) The recent development of precision agriculture hardware: global positioning system (GPS)-guided agricultural machinery for applying agrochemicals and recording harvests in a site-specific manner is now widely available. Remote-sensing and advanced electronic monitoring techniques can automatically record crop and soil conditions in space and time. In fact, we are faced with a

very strong commercial technology push that needs to be channelled into operational systems that can really be used by farmers.

(2) Computer simulation techniques are now available for predicting development and solute transport for a wide variety of crops. Thus we no longer need to make exhaustive field experiments to fine-tune our management practices: we can use computer techniques—calibrated and validated with good field data—to explore the effects of a wide range of alternative land use scenarios.

The stakes are high: more efficient use of natural resources is not only economically attractive but also favourable for environmental quality. This desirable combination of effects is crucial for future agricultural production systems. A single and rather one-sided thrust on environmental requirements has produced much antagonism in agricultural circles. Precision agriculture can provide a breakthrough here: what is good for business is also good for the environment.

Some options for precision agriculture

Two broad approaches to precision agriculture may be distinguished at this time:

(1) A 'reactive', monitoring approach where sensors are not only used to continually monitor soil and crop conditions, but also to guide on-the-go management practices, such as tillage, fertilization, and weed and disease control.

(2) A 'proactive' process-oriented approach where predictive modelling is used to anticipate moments of stress within fields. Thus, the occurrence of stress, which is the key component of procedure (1), is avoided and so is the associated crop damage.

Procedure (1) appears to be quite attractive. As soon as differences in crop reaction are noticed within a field (e.g. by using remote-sensing techniques) corrective measures can be taken by on-the-go management techniques, using a variety of yet-to-be-developed sensors. This procedure has three basic and serious limitations:

(a) Stress phenomena in the crop have to be visible before a reaction can be given. By that time some damage is already likely. It would be preferable to avoid damage by anticipating future effects.

(b) Stress phenomena are observed but the reasons for them are unknown. They can be many and they are not always associated with soil and weather conditions. Field observations are needed to analyse observed stress phenomena in order to prescribe corrective measures. This takes time and substantial damage may have already occurred by the time corrective measures can be implemented.

(c) Environmental side-effects of agricultural practices, defined in terms of whether or not threshold values are exceeded, are difficult to document. Even though

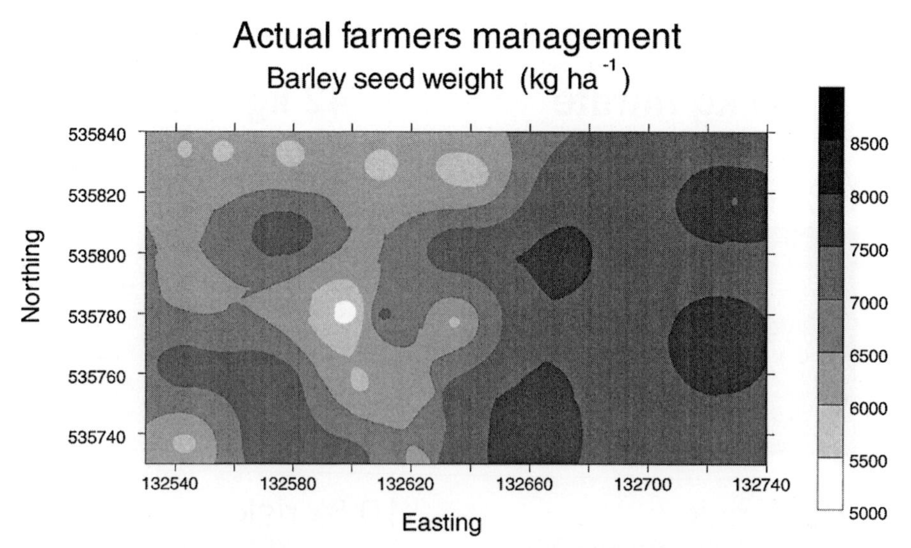

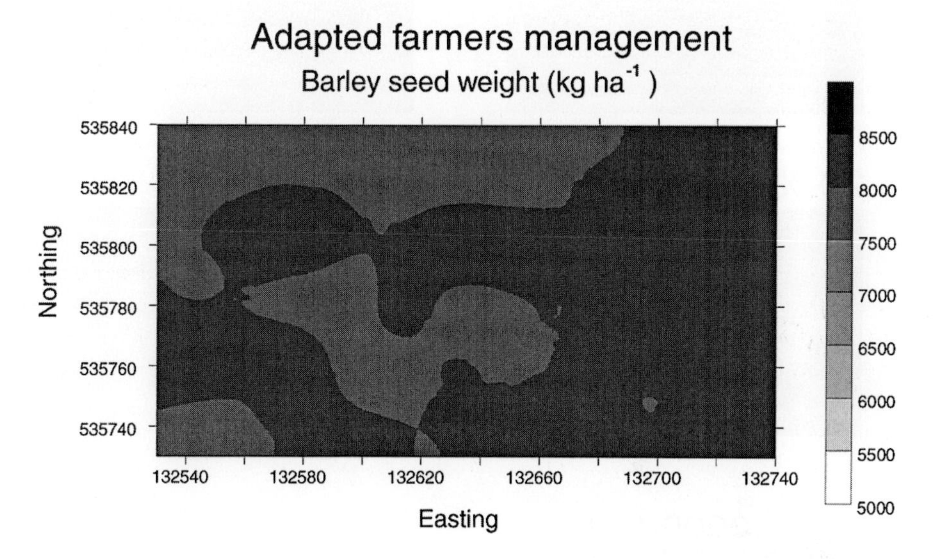

FIG. 1. Simulated barley yields based on farmers management practices in 1994 (upper part) and 20% higher yields (lower part), which would have been realized by a second nitrogen application three weeks earlier. Using a weather generator, the model had indicated a probable early N shortage. Data were calculated for 64 points that were interpolated to obtain the observed patterns, visualized with geographical information systems (GIS). From Booltink & Verhagen (1997).

35 kg nitrate

0 % risk

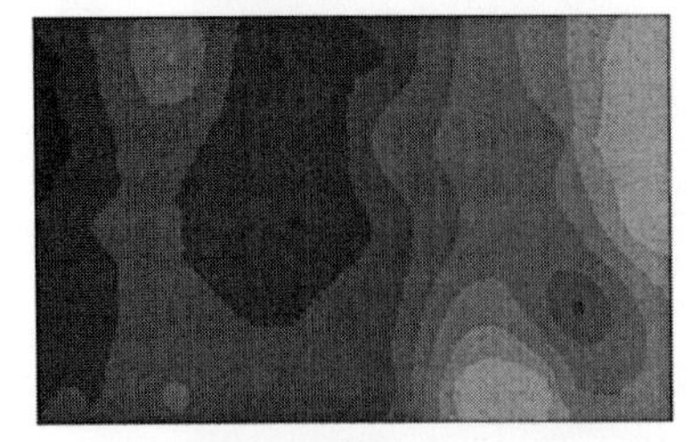

10 % risk

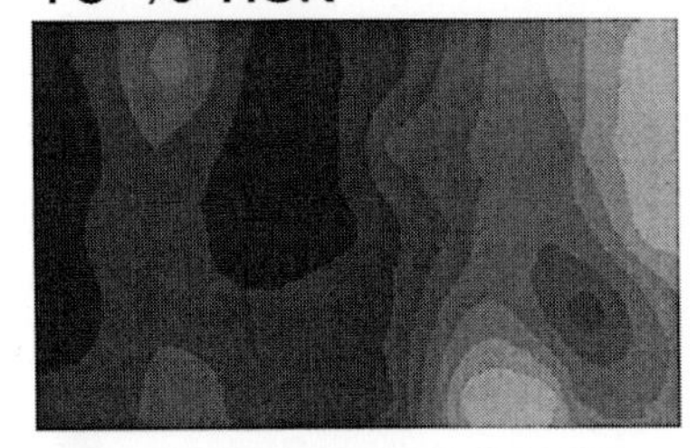

20 % risk

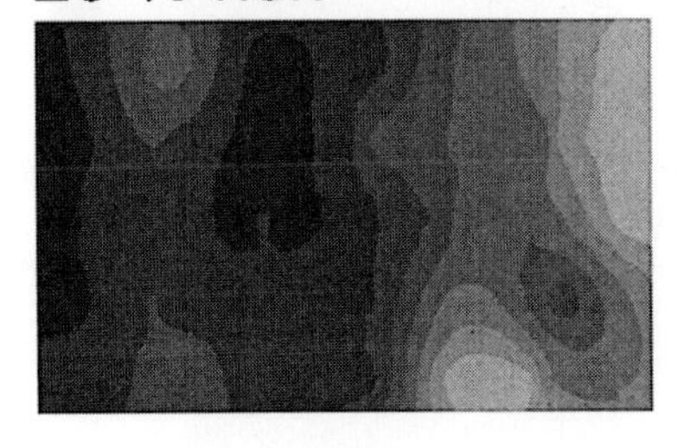

42 kg nitrate

0 % risk

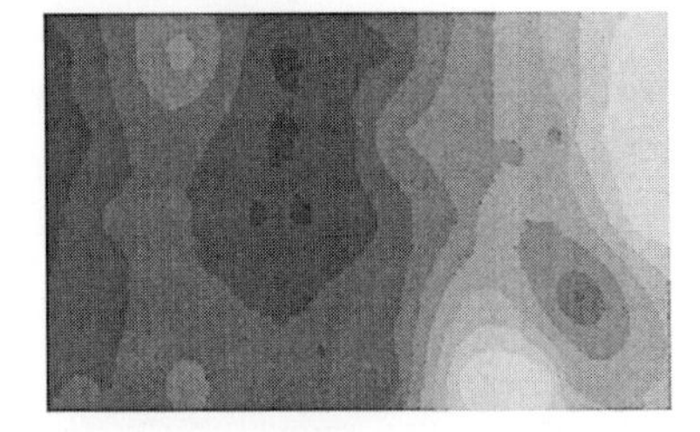

10 % risk

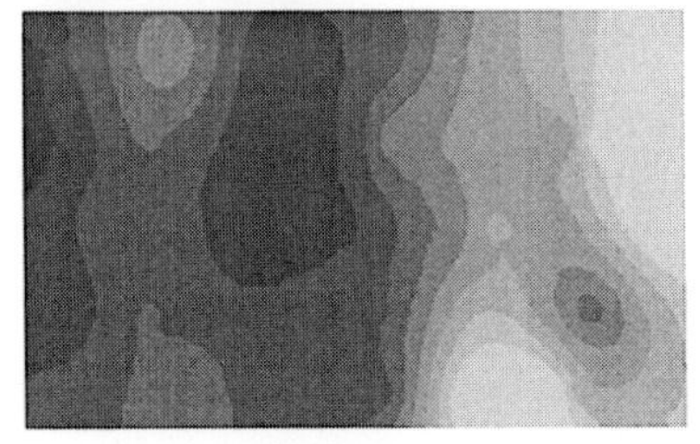

20 % risk

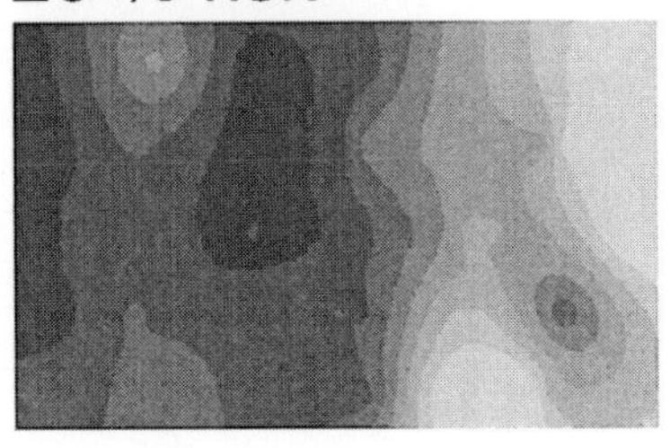

Legend

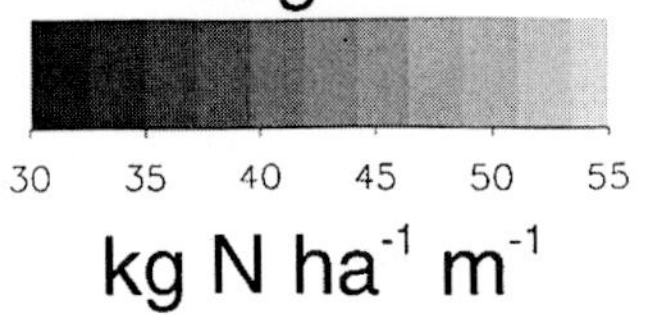

kg N ha^{-1} m^{-1}

nitrate sensors can be placed in the groundwater body, there will be no data on solute transport processes. Cumbersome field measurements may be needed.

The more proactive procedure (2) uses computer simulation models to estimate crop growth and solute fluxes. A central role is to be played here by simulations of the soil water regime. By now, such simulations are certainly possible, using calibrated and validated simulation models of varying degrees of complexity. Weather generators can be used to predict future weather conditions in the growing season and the corresponding crop growth and nutrient leaching rates (e.g. Booltink & Verhagen 1997). These authors demonstrated for a Dutch case study that a nitrogen deficit was recognized by the model and that the second fertilization, as applied by the farmer, should have been applied three weeks earlier. Then, a substantial production rise of 20% was predicted by the model (Fig. 1). Anticipation of a problem, be it a nutrient or a water shortage, and taking timely action avoids loss of production. This is a clear advantage over procedure (1) (see point (a) above). By running a model we can diagnose at least some of the problems that may lead to reductions in crop growth. For instance, water deficits and nitrogen shortages can be well predicted. They are very important factors reducing crop growth and the model will alert the user in time without the need for additional field monitoring (see point (b) above). Finally, a simulation model allows estimates of both crop growth and nutrient fluxes for a wide variety of weather conditions in a given agroecological zone. Thus, a record is provided which allows definition of realistic trade-offs between production requirements on the one hand and the need to stay below environmental threshold values on the other. Using the simulation approach, Booltink & Verhagen (1997) and Verhagen & Bouma (1997) showed that leaching of nitrates in The Netherlands occurs mainly in the wet winter season. Verhagen & Bouma (1997) also defined critical threshold values for total nitrogen in the soil at the start of the wet season, which were based on simulations for a 30 year period, allowing statistical expressions for leaching probabilities (Fig. 2). In this way, variability in space and time was expressed in terms of probabilities that a certain threshold value is exceeded. The user, regulator or politician can then make a choice as to the risk they are willing to take.

Procedure (1) would never allow these types of analyses and, more importantly, the type of documentation thus being provided on environmental quality (see point (c) above).

FIG. 2 (*opposite*) Maximum allowable total nitrogen contents at the beginning of the wet season on September 15, which are associated with nitrate leaching in the wet winter season at or below the average Dutch threshold value of 35 kg N (same field as in Fig. 1). These contents are not expressed as fixed values for each location in the field but in terms of probabilities (risks) that the threshold value may be exceeded. Data were based on simulations for a 30 year period, allowing an expression for the effect of weather variability. The three figures for the local threshold value of 42 kg N in the northern part of the country illustrate the major effect of small differences in threshold values. From Verhagen & Bouma (1997).

An outline of a future decision support system for precision agriculture

Even though research is still in progress all over the world, a broad outline of future decision support systems for precision agriculture can be sketched on the basis of experiences so far. A combination of the procedures (1) and (2) would appear to be most advantageous. To allow running of simulation models, a soil database of the farm to be characterized needs to be assembled (see e.g. Verhagen et al 1995). Measurement of the basic physical and chemical characteristics of the soil is needed, unless they can reliably be derived from pedotransfer functions.

Interpolation techniques are used to transform point calculations of crop growth and solute fluxes to areas of land. Various geostatistical techniques, such as indicator kriging, are suitable (e.g. Finke 1993, Verhagen et al 1995). Weather generators can be used to predict soil and crop conditions (e.g. Booltink & Verhagen 1997). Remote sensing data, when available, are most useful to calibrate and validate simulation models for crop growth. When either irrigation, fertilization or application of biocides is considered to be relevant in view of exploratory simulation results, on-the-go equipment can be put into action in due time before water and nutrient shortages or pests and diseases can arise. The equipment should be calibrated to ensure that environmental threshold values are not exceeded during the application of management procedures. Here, again, simulation runs are necessary to quantify the environmental impact of management measures.

Rather than run a model all over again for each point observation at any farm, we would advocate stratification of data according to soil survey concepts, particularly in relation to the soil series. This represents one particular well defined type of soil in terms of its location in the landscape and its composition. Stratifying results of simulations by soil series could yield much information that would be useable when only the soil series is established at any new farm to be surveyed, rather than a series of detailed observations. Such a procedure, which represents establishment of a meta database might be useful for a first 'quick and dirty' approach, when providing soil data for precision agriculture (e.g. Droogers & Bouma 1997).

As stated before, variability of soil and crop conditions in space and time is a central concept for precision agriculture. In the next section I will briefly review the sorts of questions that may be raised in the context of the design of decision support systems.

Questions to be answered

(1) How can we build a commercially feasible field soil database that adequately characterizes variability in space?

(2) Which soil data should we sample and how? How can the available data be used most efficiently?

(3) How can we characterize variability in time by monitoring and modelling? Which monitoring techniques should be applied? Which models should be used to

characterize crop growth and nutrient fluxes? How can we calibrate and validate the models?

(4) How should we use available know-how and new data in operational decision support systems that can be used by farmers to support their strategic, tactical and operational decision making? How can we guide complex flows of information? What are the roles of interactive decision support systems?

(5) How should we integrate spatial and temporal information of different sources in one operational georeferenced database and decision support system? What is the role of existing soil survey expertise, particularly the use of the soil series concept for extrapolating information in a 'quick and dirty' procedure?

(6) How should we define and use environmental threshold values in the context of precision agriculture that can contribute towards defining sustainable land-use systems?

Acknowledgements

Input by J. Verhagen, H.W.G. Booltink and P. Droogers is gratefully acknowledged. Part of the research reported was funded by the EU-AIR project 921204 ('IN SPACE'): 'Reduced fertilizer input by an integrated location specific monitoring and application system', coordinated by D. Goense, Wageningen, The Netherlands.

References

Booltink HWG, Verhagen J 1997 Using decision support systems to optimize barley management on spatial variable soil. In: Kropff MJ, Teng PS, Aggerwal PK, Bouma J, Bouman BAM, Jones JW, van Laar HH (eds) Applications of systems approaches at the field level. Kluwer Academic Publishers, Dordrecht, The Netherlands, p 219–235

Bouma J, Brouwer J, Verhagen J, Booltink HWG 1995 Site-specific management on field level: high and low tech approaches. In: Bouma J, Kuyvenhoven A, Bouman B, Luyten J, Zandstra H (eds) Eco regional approaches for sustainable land use and food production. Kluwer Academic, Dordrecht, The Netherlands, p 453–475

Bouma J, Verhagen J, Brouwer J, Powell JM 1997 Using systems approaches for targeting site-specific management on field level. In: Kropff MJ, Teng PS, Aggerwal PK, Bouma J, Bouman BAM, Jones JW, van Laar HH (eds) Applications of systems approaches at the field level. Kluwer Academic Publishers, Dordrecht, The Netherlands, p 25–37

Brouwer J, Bouma J 1997 Soil and crop growth variability in the Sahel. A summary of five years of research at ICRISATSC, Niamey, Niger. Information Bulletin 49. ICRISAT. Sahelian Center and Agricultural University, Wageningen

Droogers P, Bouma J 1997 Soil survey input in exploratory modeling of sustainable soil management practices. Soil Sci Soc Am J, in press

Finke PA 1993 Field scale variability of soil structure and its impact on crop growth and nitrate leaching in the analysis of fertilizing scenarios. Geoderma 60:89–109

Soil Science Society of America (SSSA)1995 SSSA statement on soil quality. Agronomy News, June 7, SSSA Madison, WI, USA

Verhagen J, Bouma J 1997 Defining threshold values at field level: a matter of space and time. Geoderma (special issue), in press

Verhagen J, Booltink HWG, Bouma J 1995 Site-specific management: balancing production and environmental requirements at farm level. Agric Sys 49:369–384

DISCUSSION

Burrough: You made a very categorical statement about modelling: you said that we should model first and interpolate second. This might be appropriate when you're modelling point processes, but if you're also modelling spatial interactions you have to take those into account before you generalize. For instance, this would be the case with hydrological modelling, run-off, surface water transport or leaching. It is something we should bear in mind — it's not just modelling at a point and then interpolating the results. Rather, how you do the modelling depends on the kinds of processes that you're dealing with. I know that Jaime Gómez-Hernández is doing a lot of hydrological modelling in which he uses geostatistics and conditional simulation to assign values to the model blocks to fully support the transport functions. In some nitrate work we are doing on the Rhine catchment we have seen that putting transport functions into the modelling gives us enormous improvement in the results with the validation data (M. de Wit, personal communication 1997).

Bouma: That is a good point. We should bear in mind that in Holland the land is flat, and I have referred to process modelling in a flat field. Only some 15% of the surface of the Earth is flat; the rest is hilly. Here, modelling of landscape processes is crucial and point data cannot be treated separately.

Bregt: You said that temporal variation may be more important than spatial variation for the application of precision agriculture. Isn't this a big problem for precision agriculture in the sense that the dominant source of temporal variation — the weather — is inherently unpredictable? If you want to apply fertilizer, then for your decision making you will want to know the weather conditions for the coming season.

Bouma: I said that the variation in time is more important in our case study than the variation in space. This is a rather blunt statement. It varies, of course, depending on the agroecological zone you are in and the soil characteristics. What strikes me is that even in Holland, which has a rather regular oceanic climate, the temporal variation among the seasons is quite large. Many countries have more extreme climates, so I would expect the differences there to be even greater. I agree with you that temporal variation is one of the big challenges facing farmers in their decision-making process. This 'looking ahead' aspect is something that we haven't been good at: we have done a lot of modelling looking backwards but relatively little looking forwards.

Stein: In your modelling you have used a weather generator. How reliable is this and how can it be used? For example, how far forwards can it project?

Bouma: I refer to the work of Booltink & Verhagen (1997), who used the DSSAT weather generator, allowing predictions of about two week periods. Meteorologists expect that within a few years they will be able to predict the weather for two to three week periods.

McBratney: In Australia it is certainly the case that weather forecasting has been improving over the last 10 years since meteorologists began to understand the Southern oscillation index and the El Ninõ effect.

Burrough: In Northern Europe there are fewer sea-based stations, which is part of the problem.

McBratney: I agree entirely with Arnold Bregt that the major problem here is weather prediction. Is there an interaction between the weather and the soil environment? If there is, then you can probably sensibly divide up fields into smaller regions, but if there isn't, it's going to be rather difficult to take advantage of this soil variability. In the latter case, the traditional approach of uniform management might be a reasonable one under such a high-risk situation.

Meshalkina: I would like to ask a question about the change of the spatial variability in time. What do you do if the same location in space belongs to different spatial patterns at different times?

Bouma: Each point in time has a certain pattern which we obtain by interpolation of point data. Patterns (of yields and leaching, for instance) change from one time to the next. We have applied map comparison techniques to derive patterns with a general validity. The sandy and the clayey spots come out all the time. In a wet year the sandy spots have higher yields than the clayey spots; in a dry year it is the other way round. But the overall pattern is the same.

Barnett: I would like to return to this question of the difference between temporal and spatial variation. I may have misinterpreted your paper, but were you implying that you believe that temporal variation is more important than spatial variation?

Bouma: That is my impression. For our case study, when you look at a field over several years, the temporal variation has at least as much impact on crop growth and leaching as spatial variation does. This statement, of course, has no general validity because soil and weather variability are quite different in different agroecological zones.

Barnett: Surely that depends entirely on what variables you're interested in measuring, and what the scale of those variables is. It is possible to construct examples where the temporal variation could be of no significance at all in terms of certain types of variables measured on certain scales. On the other hand it could be totally overwhelming. I think it is a rather sweeping generalization.

McBratney: Can you be more specific? What sort of variables are you talking about?

Barnett: You can think of trivial examples where temporal variation is going to be insignificant. For instance, if you are measuring the potassium level in the soil at the same point day after day, the temporal variation is negligible compared with the spatial variation over the field.

Bouma: The same is true for texture, of course.

Barnett: One of the greatest demands we have before us is to find adequate ways of harmonizing, within models and methods, spatial and temporal variation. At the moment we have a combination of separate spatial methods and temporal methods which we try to patch together, but we have not had much success in introducing unified approaches.

Rabbinge: You began by mentioning that farmers have decisions to make on three different levels: strategic, tactical and operational. The majority of current discussions in precision agriculture are at the tactical and operational levels. However, broader strategic decisions concerning land use can also be dictated by the characteristics and heterogeneity of a particular field. Have you thought how this technology could be developed to influence these sorts of strategic decisions?

Bouma: What is actually done at the field level is decided only after a lot of choices have already been made, because a farmer has many things to consider. There is, as you say, a risk that we concentrate so much on the operational details that we forget about the tactical and strategic ones. Ultimately, we want a decision support system for the farmer in which spatiotemporal heterogeneity plays its proper role. There are many decisions to be taken in terms of doing things at different times. There are peak times during which the farmers may be unable to make interventions at the ideal time because they have too much else to do. There are trade-offs to be made. But choosing precision agriculture certainly represents a strategic decision with many implications.

McBratney: At the sort of scale that we're working at, which is several thousand hectares per farm, in the first instance the practical outcome of all of the yield monitoring and soil monitoring work we're doing will be to make the sort of strategic decisions that we have talked about—what to grow in which field. The main mechanism by which we do this is to turn our yields into dollars and make decisions based on the futures market. These decisions will be made on the basis of this high level of information: much higher than we would have had in the past. In the first instance that will have a much bigger impact than the differential management within fields.

Robert: In the Midwest region of the United States, it has been found that conventional fertilizer recommendations generally don't perform well with site-specific crop management. They certainly don't optimize the potential of precise management. Present recommendations were developed from research on a few agricultural experimental stations and limited on-farm plot experiments, often located where natural conditions were as uniform as possible. They rarely represent the broad range of variability that exists on farms resulting from variable natural conditions and pest management practices. For this reason, it is recommended that producers should install a yield monitor on combines and start simple experiments on fields or portions of fields representative of the type of variability occurring on the farm. The experimental design uses field length strips crossing common variable soil or site conditions. The selected practice stays constant within each strip but treatments vary between strips and are replicated several times within the same field or on other fields. These experiments can be performed with the help of local agricultural consultants, extension agents and university researchers.

Rabbinge: Another point I would like to raise is the very last point Johan Bouma made, concerning motives for adopting precision agriculture. It can be done for reasons of poverty and for reasons of wealth. The poverty motive is illustrated by the example from Niger where farmers use scarce resources in the most appropriate way in order to have the maximum output per unit of input. The wealth motive is trying to limit over-use of external inputs that will lead to negative environmental side-effects. These are the main reasons for using precision agricultural techniques.

Webster: The small farmer in Africa who cultivates by hand is already a much more precise farmer than his European or North American counterpart. What the subsistence farmer does is to reinforce success: he concentrates where his input is likely to have the biggest single output. It is a situation that is very different from that we face in Western Europe, where the risk is concerned with the environment — it is not the risk of starvation.

Mulla: I would like to ask about the management of environmental risks. Let me give a specific example of the management of nitrate leaching on sandy soils versus fine-textured soils. A typical producer would be interested in maximizing their productivity in general. How do you deal with the situation of managing a sandy portion of a field where there is a high risk of nitrate leaching? The farmer may think that this is because the nitrate levels in that part of the field are low and that applying more nitrogen is the solution to getting better productivity. How do you convince a farmer that higher productivity is not necessarily the objective, but that managing for environmental quality is the goal? Applying either lower rates of nitrogen or applying rates of nitrogen at the right time of the year is a better strategy than just over-applying in order to compensate for leaching.

Bouma: That's a crucial issue. In Europe, environmental regulations are being tightened, and will continue to become more stringent. One of the great concerns we have is that these rules and regulations are increasingly being written by lawyers and economists, who may not have a thorough grasp of the real processes that take place. The rules are very rigid. We need to get in on that decision-making process. Let's come with examples and let's show the risks of making wrong assessments. Then we can ask: what risk level is society willing to tolerate?

The second point concerns how we convince the farmer. There is this cliché of the 'win/win' scenario, but it is a good one. I see many instances in precision farming where this really is the case: precision agriculture increases the use-efficiency of fertilizers, thereby cutting costs, while reducing or avoiding leaching losses! Precision agriculture presents many possibilities because it sits well with the gut feelings of the farmers. So here are the two lines of activity: we need to get in touch with the regulators (show and tell, and raise awareness), and with the farmers (trying to get the 'win/win' situation).

Reference

Booltink HWG, Verhagen J 1997 Using decision support systems to optimize barley management on spatial variable soil. In: Kropff MJ, Teng PS, Aggerwal PK, Bouma J, Bouman BAM, Jones JW, van Laar HH (eds) Applications of systems approaches at the field level. Kluwer Academic Publishers, Dordrecht, The Netherlands, p 219–235

Spatial variability of soil moisture regimes at different scales: implications in the context of precision agriculture

Marc Voltz

Laboratoire de Science du Sol, INRA, 2 Place Viala, 34060 Montpellier Cedex 1, France

Abstract. Precision agriculture is based on the concept of soil-specific management, which aims to adapt management within a field according to specific site conditions in order to maximize production and minimize environmental damage. This paper examines how the nature and sources of variation in soil moisture regimes affect our ability to simulate soil water behaviour within a field with adequate precision in order to advise optimal soil-specific management. Field examples of variation in soil moisture regimes are described to illustrate the difficulties involved. A discussion identifies three main points. First, it is recognized that the current modelling approaches to soil moisture regimes do not sufficiently account for local heterogeneities in soil and crop characteristics such as soil morphology and rooting patterns. Second, the estimation of within-field variation of soil hydraulic properties is difficult because of large short-range variation of the properties and general lack of observed data; one way to overcome this problem is to seek new measurement techniques or to find easy-to-measure auxiliary variables spatially correlated to the variables of interest. Last, as pollution impacts often become noticeable to society at scales larger than the scale of agricultural management, hydrological modelling can serve for linking both scales and advising agricultural practices that minimize undesirable pollution effects.

1997 Precision agriculture: spatial and temporal variability of environmental quality. Wiley, Chichester (Ciba Foundation Symposium 210) p 18–37

Precision agriculture is based on the concept of soil-specific management, which aims at adapting management within a field according to specific site conditions in order to maximize production and minimize environmental damage. This is now possible because technology is emerging that enables soil management to be altered automatically as equipment traverses the field (Robert 1993). However, soil-specific management will be relevant only if agronomists are able to deliver accurate site-specific advice. This requires precise information on specific soil conditions, especially soil characteristics and soil behaviour. Among the latter, soil moisture regimes are certainly an essential piece of information: soil moisture exerts a critical control on crop development and pollution phenomena. However, soil moisture regimes are largely variable in space and time because of variable soil properties and variable soil boundary conditions. As it is not possible to measure them continuously at all points within a field, prediction by simulation is necessary. Consequently, it must be asked whether present modelling approaches are capable of simulating soil moisture regimes with adequate precision to give the farmer precise advice for site-specific management. This paper examines this question with regard to the nature and sources of spatial variation in soil moisture regimes. In the first section, scales of interest to precision agriculture are defined. A second section presents field examples of variation at different scales of soil moisture regimes and of surface water pollution. A third section discusses some issues related to the simulation of soil moisture regimes in the context of soil-specific management.

Scales of spatial variation

The word 'scale' refers here to the size of the area under consideration. In an agricultural context three scales are of interest. The first is the management scale, which traditionally corresponded to the field scale and which comes down now to the pedon scale, i.e. a few square meters, in the case of soil-specific management practices. Second is the scale of the farm. This is a less definable scale because it consists of a set of fields that are not necessarily adjacent. However, with regard to the proportion of landscape heterogeneity that the farm scale incorporates, it may be likened to the catchment scale in many instances. Last is the scale at which the environmental impact of agricultural practices is assessed. This scale varies according to the environmental problem that is studied. For soil pollution phenomena it can be similar to the management scale; for water pollution the scale is often much larger and may correspond to the scale of a groundwater reservoir or of a catchment.

Case studies

In this section, in order to illustrate the problems involved with describing and modelling the spatial variability of soil moisture regimes at the scales of interest to precision agriculture, I describe examples from a cultivated catchment in Southern France (Voltz & Andrieux 1995). The catchment is 91 ha large (see Fig. 1), and is located in the Hérault department, in the commune of Roujan (43°30′ N, 3°19′ E). The main crop is grapes. The site is primarily man-made, with terraced slopes and a major network of ditches collecting the run-off water. In geological terms, the substrata are derived from marine, lacustrine or fluvial sediments. As shown in Fig. 2, four topographic sections are distinguished within the catchment: a plateau in the upper part, terraces at midslope, a glacis and a depression. According to the F.A.O. soil classification system, the main soil types are luvisols, regosols and cambisols. The climate is sub-humid Mediterranean, with a prolonged dry summer period. Average

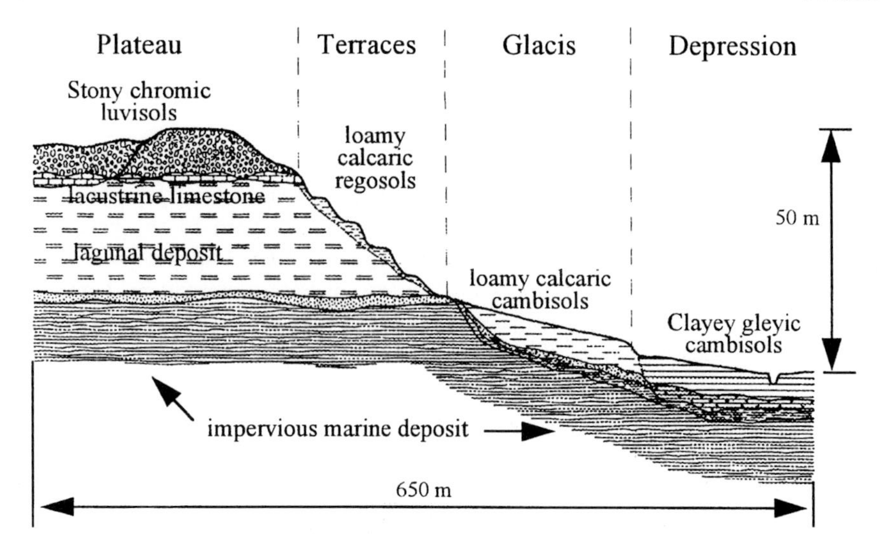

FIG. 1. Transverse section of the Roujan catchment.

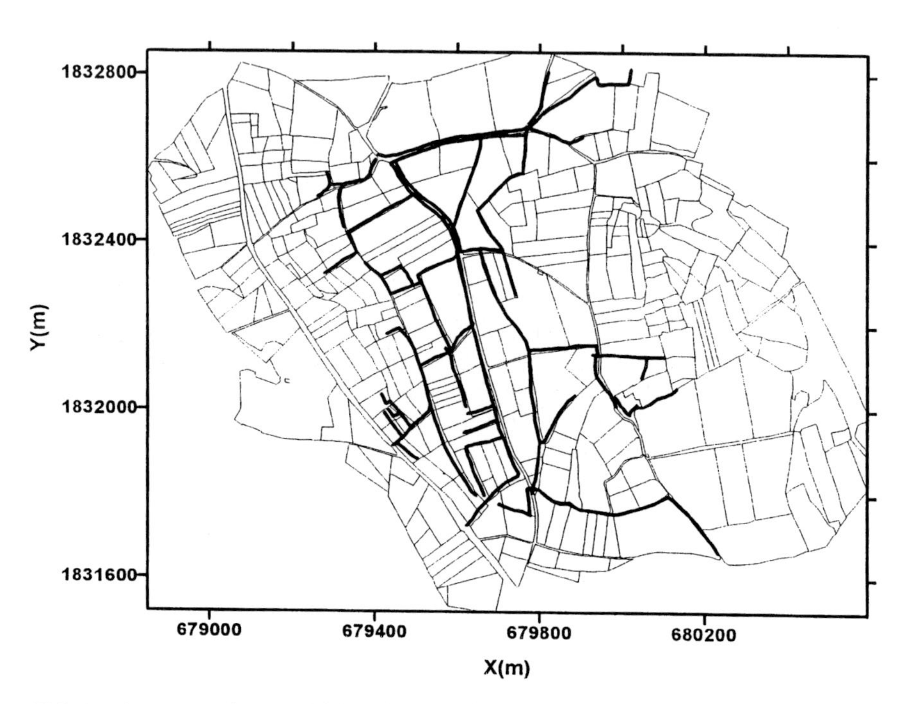

FIG. 2. Roujan catchment with location of field boundaries (thin lines) and network of ditches (bold lines). The X and Y axes correspond to Lambert zone II coordinates.

annual rainfall is 650 mm and average annual Penman potential evapotranspiration is 1090 mm. Rainfall mainly occurs as storms.

Variation in soil moisture regimes at the field scale

Many studies have considered the spatial variation in soil hydraulic properties to be an important factor of the spatial variation in soil moisture regimes at the field scale (e.g. Peck et al 1977, Bresler & Dagan 1983, Braud et al 1995). Hereafter, work by Trambouze (1996) is used to show the influence of another factor, i.e. the heterogeneity of the rooting system of crops. Trambouze (1996) analysed the spatial variation in soil water storage of a 1 ha vineyard, in which grapevines are grown in rows. The row spacing was 2.5 m and vinestock spacing along each row was 1.25 m. Figure 3 shows the spatial variation in soil water storage changes during summer 1993 between five plots within the vineyard, and between row and inter-row positions within each plot. Two scales of variation in soil water storage changes can be distinguished. The first is the row/inter-row scale: for all plots changes in water storage were less under the inter-row than under the row. This arose from the decrease in rooting density and rooting depth according to the distance from the row. The second is the field scale and is superimposed on the former: marked differences between plots exist. A detailed analysis of the profiles of soil water content revealed that the differences in water storage changes between the plots were also due to differences in the depth of water extraction by roots. By the observation of

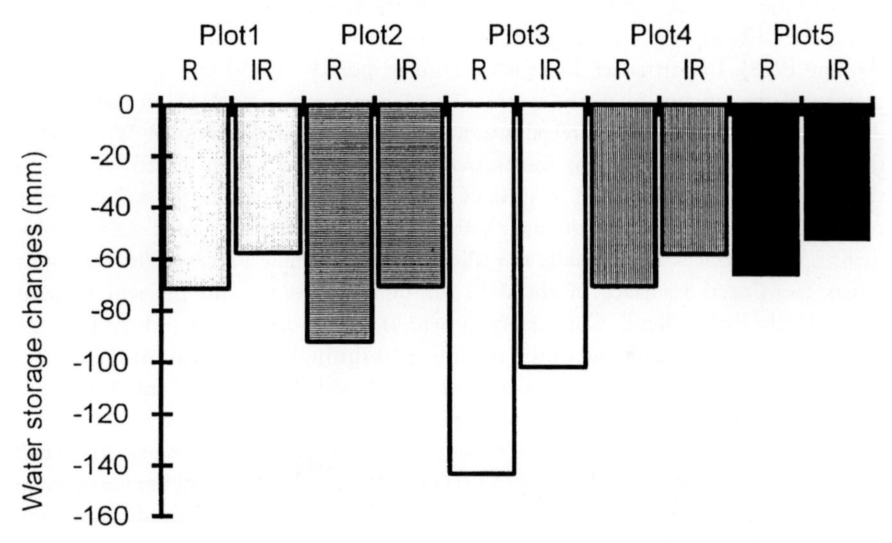

FIG. 3. Spatial variation in soil water storage in a vineyard during summer 1993 (from 2 July to 3 Sept) (after Trambouze 1996). R and IR distinguish the measurement positions, row and inter-row, within each of the five experimental plots.

soil profiles, the latter differences could be related to the presence of a compacted layer which forms a barrier to roots and whose depth varies irregularly within the field. The study of Trambouze (1996) is an example where the major factor of variation in a soil moisture regime is the spatial distribution of roots as related to the planting geometry and to the variation in compactness of the soil. It illustrates two main difficulties for representing the spatial variation in soil moisture regimes at the field scale. First is the difficulty of simulating the average soil moisture regimes at the scale of the management unit. In a vineyard the management units are the vinestocks and their rooting zones. Current models of soil moisture regimes ignore the spatial heterogeneity of rooting systems and only consider average root densities at the modelling scale. This is clearly a drawback for sparse crops but it can also be so for uniform crops. Tardieu et al (1992) showed by a simulation exercise that measured patterns of the rooting system of a maize crop whether represented in an organized way or by their average characteristics leads to different values of simulated evapotranspiration. The second difficulty lies in estimating the distribution of the parameters of the modelling approach over the whole field. This difficulty is general, but is increased in the above example because the main parameter is the rooting density which is particularly arduous to measure and, therefore, to map.

Variation in soil moisture regimes at the catchment scale

In this example, I illustrate the influence of the spatial variation in boundary conditions on local soil water regimes. Figure 4 shows the difference in soil water dynamics between the four sections of the Roujan catchment. The variation in soil moisture along the hill slope is very different from that already observed in humid regions (Dunne 1978). In particular, there is no regular increase in soil saturation from the plateau of the catchment to the depression. A water table develops on the plateau during the rainy periods and recedes to a depth of 5 m during summer. Water from the plateau is mainly caught by the network of ditches and is thus directly routed to the depression. Consequently, in spite of their mid-slope position, the soil moisture regime of the terraces is almost solely influenced by intercepted rainfall. In addition, fields on the terraces are generally not tilled because of limited accessibility, and thus crusts form at the surface of the soil, run-off is increased and the soil remains permanently dry at depth. Conversely, downhill, in the depression and the glacis, a general water table exists due to the reinfiltration through the bottom of the ditches of the water drained from the plateau and the terraces. The soils of the depression remain close to saturation throughout the year. The large scale variation in soil moisture regimes that exists at the level of the Roujan catchment subsumes the small scale variation described in the previous section. It arises largely from the variation in soil boundary conditions, which is caused by the lateral redistribution of surface and subsurface water within the catchment.

This example illustrates the fact that, for predicting local moisture regimes within the fields of a farm, large-scale sources of variation must also be considered. Among

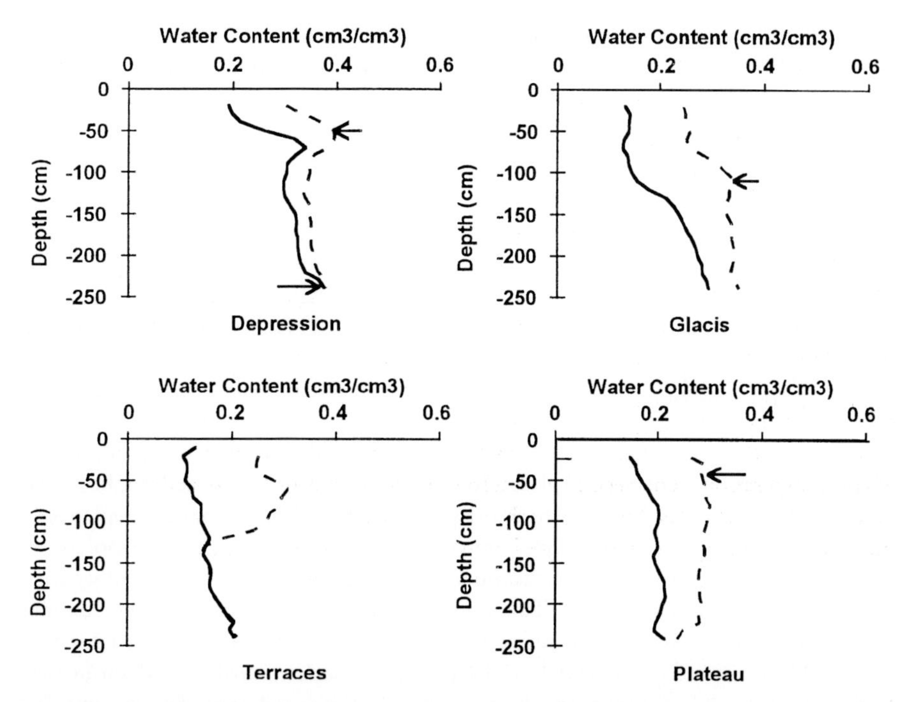

FIG. 4. Minimum and maximum observed water content profiles in the four catchment sections in 1993. The arrows indicate the depth of the water table at the time of measurement of the profiles.

them topography and human-induced modifications of landscape characteristics (e.g. network of ditches, soil surface features) are important because they directly influence surface and subsurface flow.

Variation in surface water pollution according to the scale of analysis

This example is taken from an unpublished paper by B. Lennartz, X. Louchart, P. Andrieux & M. Voltz (6th Workshop Pesticide in Soil & Environ, 13–15 May 1996). These authors monitored losses of herbicides by run-off water at the field scale and at the outlet of the Roujan catchment. Two kinds of fields were distinguished according to the agricultural practices in use for controlling weeds. A first practice is to cultivate the soil between the rows of vinestocks and to spray herbicides along the rows. The second practice consists of spraying herbicides over the whole vineyard without any tillage. Table 1 summarizes the results that were obtained in 1995. Variation in herbicide losses at the field outlets according to cultivation practice is moderate and mainly due to less overland flow on the tilled field than on the untilled field. In contrast, a large difference in losses is observed between the field scale and the

TABLE 1 Losses in herbicides (given as percentage of herbicide applied) at the outlets of tilled and untilled fields and at the outlet of the Roujan catchment in 1995

	Scale of analysis		
	Untilled field	*Tilled field*	*Catchment*
Diuron	2.1	0.47	0.07
Simazine	0.2	0.87	0.03

After B. Lennartz, X. Louchart, P. Andrieux and M. Voltz (unpublished paper, 6th Workshop Pesticide in Soil & Environ, 13–15 May 1996).
Applied quantities at the catchment scale were estimated by a survey and by assuming that on the tilled fields only one-third of the area was sprayed.

catchment scale. This can be explained as follows. A part of the overland flow that occurs on the fields reinfiltrates in the depression of the catchment and replenishes the water table. Consequently, a proportion of the herbicides that are transported by run-off water from the field edges is retained in the depression of the catchment where it can be adsorbed and degraded. This phenomenon is maximal during the summer period when soils are dry and water tables are low. This example illustrates two points. First, pollution intensity may vary with the scale of analysis, which in turn means that it is necessary to define which scale is relevant for quantifying environmental risk. Second, to devise agricultural management practices minimizing environmental risk, it is necessary to link processes at the management scale to processes at the scale of diagnosis.

Discussion

The examples described above show that factors of variation in soil moisture regimes exist at different scales and are diverse: rooting patterns, soil properties, boundary conditions and tillage practices. This raises three kinds of difficulties in the context of soil-specific management: (1) modelling soil moisture regimes at the management scale, (2) parameterizing the models at the farm scale, and (3) linking the local soil behaviour and management practices to their environmental impact. Before discussing the possibilities to overcome the difficulties, it is necessary to define more precisely the target, that is, the precision that should be attained when simulating at the management scale. Basically, if we want site-specific management to be more effective than conventional farming, we must be able to differentiate soil properties and soil behaviour between the management sites of a given field. The precision of the simulation of local soil moisture regimes must therefore be sufficient for that purpose. Given the uncertainties that can affect the simulation of soil moisture regimes, we can expect that the above constraint will not be met easily, and will require a number of improvements in the modelling methodologies.

Modelling approaches to soil moisture regimes at the management scale

In the modelling of soil water regimes, as well as soil behaviour in general, one basic question concerns which modelling approach will provide the most accurate results. The debate is often about whether or not it is better to use simple models with limited data demand than sophisticated models (Wagenet & Bouma 1993, Addiscott 1993, Leenhardt et al 1995). There is no general answer to this question but it is interesting to examine it in the context of soil-specific management.

The distinction between sophisticated and simple models of soil moisture regimes corresponds broadly to the distinction made by Addiscott & Wagenet (1985) between mechanistic models and functional models. Mechanistic models attempt to describe the most fundamental mechanisms of flow processes whereas functional models represent the processes empirically or by using analogies. Another major characteristic of the mechanistic models is that they require much more input information than the functional models. In principle, model bias should be smaller with mechanistic models than with functional models because the latter simplify the representation of the actual processes more than the former. But Addiscott (1993) reported that functional models often give simulations that are at least as good as those of mechanistic models. One reason that is invoked is that models of water flow or soil behaviour are often used at scales much larger than the scale at which they were developed. Then the so-called mechanistic models cease to be fully mechanistic (Beven 1989, Addiscott 1993). This was true in the context of conventional agriculture in which models are applied at the field scale. But, in the context of soil-specific management, the scale of application is almost similar to the scale at which most models of soil moisture regimes are developed. Another reason is that most models are built with an assumption of homogeneity with respect to soil parameters and boundary conditions. It is now recognized that this is a gross approximation. It is to be expected, therefore, that the advantage of mechanistic approaches over other approaches will not accrue unless the effects of critical sources of variability are taken into account. Accordingly, Bouma (1993) recommended building models that are based on expressions for heterogeneity. In this respect several sources of heterogeneity that are not usually taken into account seem of importance for improving the simulation of soil moisture regimes on farmed land: soil morphology (e.g. Bouma 1984, Booltink 1993), tillage practices (e.g. Van Es 1993) and rooting patterns (as shown in the example above).

Predicting variation in model parameters and boundary conditions at the farm scale

Whichever modelling approach is used, simulating soil moisture regimes at the farm scale requires an estimation of the distribution of the model parameters and boundary conditions within all fields of the farm. Of course, the estimates should be as precise as possible because their precision directly influences the model simulations. At least, the error variance of the estimates of model parameters and boundary conditions should be less than their within-field variance. Otherwise, predicting the latter variables at any site by their field averages would be as precise, and in turn the within-field variation in soil moisture conditions could not be predicted by simulation. Thus, procedures for estimating the model parameters and boundary conditions should be judged against the criteria of precision described above.

Two kinds of estimation procedure are classically employed. The first uses information from soil survey which classifies the soil in a region and delineates the soil classes on the ground. In general, it is assumed that soil properties vary little within each mapping unit and, consequently, that their values at any site within a mapping unit can be predicted by the mapping unit means. The precision of this procedure depends closely on the mapping resolution (Webster & Beckett 1968, Marsman & de Gruijter 1986, Leenhardt et al 1994). The optimal resolution for predicting soil properties in the context of soil-specific management should be defined with regard to the variation scales of the soilscapes (Robert 1993), but also with regard to the scale of the fields. Concerning the latter criterion, it is clear that if the average field size is much smaller than the average mapping unit size, soil survey cannot help to differentiate variation in soil properties within a field. Although the optimal resolution varies from one landscape to the other, it is to be expected that only detailed soil maps (<1:25 000) can provide estimates with sufficient precision for soil-specific management. Unfortunately, detailed maps are only available for small areas (Zinck 1990), even in countries with intensive farming systems.

The second estimation procedure uses numerical interpolation techniques, among which kriging seems to be the most reliable (e.g. Laslett et al 1987, Voltz & Webster 1990). Its precision depends both on the spatial correlation structure and on the density of observations of the variable to be interpolated. For many soil variables a large proportion of their spatial variation often occurs over short distances (e.g. Russo & Bresler 1981, Viera et al 1981, Van Es et al 1991). Thus, large sampling densities are required to obtain reliable estimates of these variables by numerical interpolation.

It should be noted that whatever the refinement of the above procedures, a basic limitation to their precision, when applied to soil hydraulic properties or to soil boundary conditions, is the general lack of data. This comes from the sampling and measurement costs of these variables. To overcome this problem, several authors (Odeh et al 1995, Knotters et al 1995, Bourennane et al 1996) have considered the possibility of correlating the expensive-to-measure variable with auxiliary soil or landform attributes whose spatial distribution can be more easily determined. The interpolation of the former variable can then be improved by using measurements of the easy-to-measure covariate. This approach seems promising, but depends on whether efficient covariates can be found. For soil hydraulic properties and other model parameters, the current approach is to seek possible covariates among basic soil data (soil texture, organic matter content, type of soil horizons) (e.g. Wösten et al 1985, Ritchie & Crum 1988, Vereecken et al 1989, 1990). For soil boundary conditions the problem is more complex because their values and their spatial structures are fluctuating with time in relation to soil water regimes. In the case of upper boundary conditions, such as potential evapotranspiration, moisture content of topsoil and soil surface features, remote sensing is certainly a valuable source of easy-to-measure covariates (e.g. Reiniger & Seguin 1986). For other boundary conditions such as water table levels or lateral flow, which can be an important source of variation in soil moisture regimes as shown in the above example, none of the previous procedures seems to be relevant. In the absence of a dense network of monitoring sites, only a hydrological modelling approach that simulates these boundary conditions may yield a solution.

Linking management scale and scale of environmental diagnoses of water quality

Water pollution by agricultural by-products is initiated at a local scale, as influenced by site-specific moisture regimes and management practices. But water pollution becomes noticeable and disturbing for society at larger scales. Consequently, pollution processes need to be controlled at the management scale in order to keep water quality above some threshold at the scale of social impact. As shown in the third case study, in some situations controlling pollution processes according to local criteria could lead only to the excessive constraint of agricultural practices. To optimally advise agricultural management practices in order to minimize the environmental risk, it is therefore necessary to analyse what effects small-scale management practices exert on larger-scale processes. For this purpose, models for upscaling processes must be sought. When water quality is under question, distributed hydrological modelling approaches should be useful. Unfortunately, there are still severe limitations to the operational use of hydrological models, among which are the difficulties related to the integration of processes at different scales, to the representation of landscape heterogeneities, and to model parameterization. Although it is beyond the scope of this paper to discuss this in detail (see e.g. Rose et al 1988, Beven 1989, Goodrich & Woolhiser 1991, Blöschl & Sivapalan 1995), it is interesting to point out some of the constraints hydrological models should meet if used for advising agricultural practices. First of all they should represent landscape heterogeneities arising from agricultural management (ditches, banks, terraces, tillage practices) in order to be able to predict hydrological changes due to changes in land management. Second, the capacity of long-term simulation of hydrological processes is needed for evaluating alternative strategies in the chronology of management practices. Third, parameter calibration should be as limited as possible to allow the use of the model for the test of new, currently non-existent management scenarios.

Conclusions

The emergence of soil-specific management techniques challenges somewhat our ability to predict local soil behaviour for advising optimal management practices. In this respect, the large variation in soil moisture regimes that exist at different spatial scales raises several issues that will need attention in the future:

(1) The current modelling approaches to soil moisture regimes all too often consider the soil to be homogeneous at the pedon scale. Consequently, important sources of variation in soil moisture regimes, such as heterogeneities in rooting patterns and in soil structure, are neglected. We need to identify which heterogeneities are important for the simulation of soil moisture, and to seek modelling approaches that can take account of these heterogeneities explicitly.
(2) For optimizing soil-specific management, within-field variation in soil hydraulic properties and soil boundary conditions must be estimated. However, in many instances the density of observations that is available is small and, thus, interpolation cannot provide reliable estimates whatever the estimation method. One way to overcome this major difficulty is to look for new measurement methods that allow dense sampling. Another is to seek easy-to-measure variables that are well correlated spatially to the soil hydraulic properties.
(3) The use of hydrological catchment modelling could be beneficial in the context of precision agriculture. First, it could help to define more precisely the hydraulic boundary conditions at the field scale, which is important for predicting accurately local soil moisture regimes. Second, as pollution impacts often become noticeable to the society at scales larger than the scale of agricultural management practices, hydrological modelling can serve for linking both scales and advising agricultural practices that minimize undesirable pollution effects. This requires us to develop hydrological modelling approaches that better incorporate the landscape features arising from the agricultural management of the land.

Acknowledgements

Most of the work done on the Roujan catchment was funded by INRA through the contract 'Valorisation et Protection des Ressources en Eau'. I am grateful to P. Andrieux, P. Bertuzzi, R. Bouzigues, B. Lennartz, X. Louchart and W. Trambouze for the examples described in this paper.

References

Addiscott TM 1993 Simulation modelling and soil behaviour. Geoderma 60:15–40
Addiscott TM, Wagenet RJ 1985 Concepts of solute leaching in soils: a review of modelling approaches. J Soil Sci 36:411–424
Beven K 1989 Changing ideas in hydrology—the case of physically based models. J Hydrol 105:157–172
Blöschl G, Sivapalan M 1995 Scale issues in hydrological modelling: a review. Hydrol Process 9:251–290
Booltink HWG 1993 Morphometric methods for simulation of water flow. PhD thesis, University of Wageningen, The Netherlands
Bouma J 1984 Using soil morphology to develop measurement methods and simulation techniques for water movement in heavy clay soils. In: Bouma J, Raats (eds) PAC water and solute movement in heavy clay soils. Institute for Land Reclamation and Improvement, Wageningen (Pub. no 37) p 298–316

Bouma J 1993 Soil behavior under field conditions: differences in perception and their effects on research. Geoderma 60:1–14

Bourennane H, King D, Chéry P, Bruand A 1996 Improving the kriging of a soil variable using slope as external drift. Eur J Soil Sci 47:473–483

Braud I, Dantas-Antonino AC, Vauclin M 1995 A stochastic approach to studying the influence of the spatial variability of soil hydraulic properties on surface fluxes, temperature and humidity. J Hydrol 165:283–310

Bresler E, Dagan G 1983 Unsaturated flow in spatially variable fields. 2. Application of water flow models to various fields. Water Resource Res 19:421–428

Dunne T 1978 Field studies of hillslope processes. In: Kirkby MJ (ed) Hillslope hydrology. Wiley, Chichester, p 227–293

Goodrich DC, Woolhiser DA 1991 Catchment hydrology. Rev Geophys (suppl) p 202–209

Knotters M, Brus DJ, Oude Voshaar JH 1995 A comparison of kriging, co-kriging and kriging combined with regression for spatial interpolation of horizon depth with censored observations. Geoderma 67:227–246

Laslett GM, McBratney AB, Pahl PJ, Hutchinson MF 1987 Comparison of several spatial prediction methods for soil pH. J Soil Sci 38:325–341

Leenhardt D, Voltz M, Bornand M, Webster R 1994 Evaluating soil maps for prediction of soil water properties. Eur J Soil Sci 45:293–301

Leenhardt D, Voltz M, Rambal S 1995 A survey of several agroclimatic soil water balance models with reference to their spatial application. Eur J Agron 4:1–14

Marsman BA, de Gruijter JJ 1986 Quality of soil maps. A comparison of survey methods in a sandy area. Netherlands Soil Survey Institute, Wageningen (Soil Survey Papers 15)

Odeh IOA, McBratney AB, Chittleborough DJ 1995 Further results on prediction of soil properties from terrain attributes: heterotopic cokriging and regression-kriging. Geoderma 67:215–226

Peck AJ, Luxmoore, Stolzy JL 1977 Effects of spatial variability of soil hydraulic properties in water budget modelling. Water Resource Res 13:348–354

Reiniger R, Seguin B 1986 Surface temperature as an indicator of evapotranspiration or soil moisture. Remote Sens Rev 1:277–310

Ritchie JT, Crum J 1988 Converting soil survey characterization data into IBSNAT crop model input. In: Bouma J, Bregt AK (eds) Land qualities in space and time. Pudoc, Wageningen (Proc ISSS Symp 22–26 August 1988) p 155–167

Robert P 1993 Characterization of soil conditions at the field level for soil specific management. Geoderma 60:57–72

Rose CW, Dickinson WT, Ghadiri H, Jorgensen S 1988 Agricultural nonpoint-source runoff and sediment yield water quality models: modeler's perspective. In: DeCoursey DG (ed) Proceedings of the international symposium on water quality modeling of agricultural nonpoint sources. 1. USDA, Logan (ARS 81) p 145–169

Russo D, Bresler E 1981 Soil hydraulic properties as stochastic process. I. An analysis of field spatial variability. Soil Sci Soc Am J 45:682–687

Tardieu F, Bruckler L, Lafolie F 1992 Root clumping may affect the root water potential and the resistance to soil–root water transport. Plant Soil 140:291–301

Trambouze W 1996 Bilan hydrique de la vigne à l'échelle de la parcelle: caractérisation et modélisation. PhD thesis, University of Agronomy, Montpellier, France

Van Es HM 1993 Evaluation of temporal, spatial and tillage-induced variability for parametrization of soil infiltration. Geoderma 60:187–199

Van Es HM, Cassel DK, Daniels RB 1991 Infiltration variability and correlations with surface soil properties for an eroded hapludult. Soil Sci Soc Am J 55:386–392

Vereecken H, Maes J, Feyen J, Darius P 1989 Estimating the soil moisture retention characteristic from texture, bulk density and carbon content. Soil Sci 148:389–403

Vereecken H, Maes J, Feyen J, Darius P 1990 Estimating unsaturated hydraulic conductivity from easily measured soil properties. Soil Sci 149:1–12

Viera SR, Nielsen DR, Biggar JW 1981 Spatial variability of field-measured infiltration rate. Soil Sci Soc Am J 45:1040–1048

Voltz M, Andrieux P (eds) 1995 Etude des flux d'eau et de polluants en milieu agricole méditerranéen viticole: le programme Allegro-Roujan. Bilan des travaux 1992–1995. French National Institute for Agricultural Research, Montpellier

Voltz M, Webster R 1990 A comparison of kriging, cubic splines and classification for predicting soil properties from sample information. J Soil Sci 41:473–490

Wagenet RJ, Bouma J 1993 Introduction to operational methods to characterize soil behaviour in space and time. Geoderma 60:vii–viii

Webster R, Beckett PHT 1968 Quality and usefulness of soil maps. Nature 219:680–682

Wösten JHM, Bouma J, Stoffelsen GH 1985 Use of soil survey data for regional soil water simulation models. Soil Sci Soc Am J 49:1238–1244

Zinck JA 1990 Soil survey: epistemology of a vital discipline. ITC Journal 4:335–350

DISCUSSION

Barnett: The question of where one should measure pollution is a vital consideration. I was recently asked by the UK Royal Commission on Environmental Pollution to advise on the incorporation of uncertainty in what might be termed the 'pollution chain'. Pollution can be found and measured across a whole spectrum of levels, starting with the actions that one might take even before the pollutant is allowed to enter the environment, through its manifestation at what you might call entry locations with obvious relevance to the example you were talking about, to its presence at contact locations where you encounter it 'in the field'. There is an enormous amount of uncertainty regarding where the standards and regulations should be placed with respect to positioning within this chain. Even if you can make rather arbitrary decisions on that, then you encounter (and again this was highlighted by your talk) all those other areas of uncertainty and variation. This may be simply intrinsic variation, measurement error, or the other elements of uncertainty, most of them being highly spatiotemporal in form.

It seems to me that we have a major problem on our hands in trying to produce standards which give recognition to this enormously wide spectrum of locations for measuring pollution and the contingent uncertainties that lie in that. The report prepared for the Royal Commission has taken great trouble to try to spell these out: I hope that in its report later in the year to the government the Commission will point out these problems and press for more people to be involved in trying to resolve them.

Voltz: I agree. In addition I would insist that depending on the scale at which we want to keep pollution below some threshold, we might put more or less constraint on agricultural practices. This has direct consequences on the profitability to the farmer.

For example, French farmers are very annoyed by these pollution constraints and are worried that because of them they may no longer be able to make a living.

Burrough: One point that you made, concerning the fact that there are things happening in the catchment but that they don't necessarily propagate all the way through to the output, is actually a general statement of not just pollution transport but also of erosion. We've done quite a lot of work with overland flow transport modelling, and we've been able to show that the sediment never leaves the catchment. Rather, it goes from one place to the next. If you do something like universal soil loss equation modelling, the soil evaporates—it goes into thin air. This is because people often ignore spatial connectivity in field studies; they treat their isolated field sites as if they were laboratory experiments forgetting that these things are all connected and you've got all kinds of sinks that occur along the way, which in your case stop the pollution getting further.

On another point, we have been looking at how errors in digital elevation models can affect the run-off directions for surface overland flow. Small errors in digital elevation models can cause enormous variations in the transport pattern. It is possible, using Jaime Gómez-Hernández's techniques, to simulate this and then to look at the range of probabilities. This means that you can set up an experiment and measure one set of sites, but when you come back to validate it that realization is not there anymore; you are measuring another realization, so you can't evaluate what you were originally doing. We're beginning to get the techniques for looking at the whole spectrum of these things, but we've a long way to go. It makes the legal aspects of pollution control extremely difficult.

Rabbinge: You have to have a very clear definition of the size and characteristics of the system that you're studying.

Burrough: You can also can determine at which point in the field you're going to get the lowest relative error. It is generally at the outlet, because within the catchment itself the relative errors are usually quite high. But of course you can then, by all sorts of means, get your levels down by the time you get to the end of the catchment.

Mulla: I would like to make a statement about some of the intricacies of basin and watershed management. In the USA a perplexing paradigm has developed around this issue. Small watersheds drain into larger watersheds, which drain into basins. Frequently, environmental problems are first noticed at a very large scale, such as in the Gulf of Mexico, which receives water from large catchment systems including the Mississippi river. In the Gulf of Mexico there is a large zone of hypoxia (a zone of low oxygen) which is thought to be the result of nitrogen that is being transported by the Mississippi river system from the agricultural areas of the Midwest. The person at the farm level doesn't recognize this problem at all. The paradigm that has resulted from this is that many of the decisions about pollution are being made at a very large scale, and at the small scale people are not aware of the importance of what they do and its large-scale consequences for the environment. As regulations have proceeded to take more and more importance, they have tended to be directed at the large scale issues. Yet to get changes in the farming system we need local decision-making, and we need people at the local level to take control of what they're doing and to be interested in what is happening. As decision-making transfers to smaller and smaller scales, we are faced with the difficulty of convincing people that they have water quality problems at the small scale. This is a difficult issue to tackle, because the parameters at the small scale are often not that bad — you don't see the tremendous water quality impacts that you would at the outlet of a major drainage basin.

This is something that we as soil scientists, geographers and agronomists need to tackle: we need to be aware of how our science can go from the small to larger scales, and vice-versa from the large scales back to the small scales.

Voltz: I agree with you. This requires multidisciplinary studies, and I would add social economists to the list of disciplines you mentioned. This is because, as you say, the farmers don't seem to be concerned with what happens at very large scales. If we want them to change their practices, we also need social economists studying the economic structure of the farms, looking at whether the agriculture practices we are advising are acceptable or not, both in social and economic terms.

Bouma: I would like to add a footnote to that, concerning the perception of pollution by society and also of the setting of threshold values for pollution (because that's what we deal with in the law). Your story emphasizes that it is necessary to make pollution levels specific in terms of scales. I agree. There may be some hotspots of pollution within fields that really are a problem, but if you are only looking at the end of the pipe these might not show up. Can we live with: 'the solution to pollution is dilution'?

Burrough: One of the problems of dealing with farmers as distinct from what's coming out of the Mississippi, is that the farmers are addressed by one government organization and set of professional organizations, whereas the river is addressed by another. So at the top level of government there is a major break between one group of people who think in a deterministic way (engineers) and another group who might be convinced to think in terms of transport of nutrients and pollutants. It is a major challenge to get these two groups together, because they do not recognize that you have these kinds of problems and that you're dealing with linked systems.

Groffman: There are problems with the question of where you evaluate the pollution from a production system: do you do it at the catchment scale or field scale? In your example, if you're a farmer who happens to be on the terrace so that your herbicides go through the drains, then you're lucky because at the catchment scale your pollution is absorbed in the drain. But if you're a farmer who is in a different position on the landscape and your herbicides don't get absorbed in the drains, then you would be in much more trouble than the first farmer. We encounter this problem frequently in the USA, where farmers commonly have a buffer zone between the property and the stream. The farmer who has a buffer zone might be better off than the farmer who doesn't, or perhaps the farmer who has a buffer zone is helping out a farmer who doesn't, because he's higher up in the catchment. Should the farmer who doesn't have the buffer zone pay the farmer who does? On a certain basis, you might want to say that a fair unit of evaluation is at the field scale, but how do you measure the environmental performance at the field scale? Can we use models to answer this question? Can we use our models as regulatory tools? I know that there are people in this room who, if they are given some input data, will be able to tell us the herbicide output as a percentage of that applied. Soil scientists hate this idea, but I would suggest that in other disciplines models are used much more aggressively than ours. For instance, with economic data, the gross domestic product is not measured, it's modelled. The use of models would give us a more solid mechanistic basis for our pollution control.

Voltz: I like your comparison with the economists, but the question is: are the economic modellers doing well?

Bouma: I support the use of models, because we are now struggling with strict environmental laws, and it is likely that they will continue to get stricter for all sorts of political reasons. There are some rather bizarre schemes going around in which farmers are required to use certain methods of book-keeping of nutrient influxes and outfluxes that are cumbersome and difficult to do. When these are analysed in a quantitative way they don't make too much sense—they are nonsense in terms of the spatiotemporal heterogeneity of fields. I think we should help the farmers because they are being victimized; they are at the end of the environmental 'decision pipe'. I have a feeling that with exploratory modelling, bearing in mind all the different parameters needed, we can come up with a range of management options that would result in farmers being able to meet criteria without having to go in for all this complex sampling and book-keeping. As scientists we have a responsibility to eliminate all the nonsensical testing that farmers are likely to be required to perform in the future.

A second angle to this is that if 'what is good for the environment is also good for business', we can catch two birds in one stroke. One of the keys to precision agriculture is that one can sometimes reduce chemical inputs significantly without any loss in yield, and at the same time have good environmental side-effects which can be documented by modelling and occasional monitoring. We are trying to move in this direction, which is more realistic than having fake 'measurements' without scientific rigour.

Rabbinge: Does it therefore make sense to introduce legislation, as has been done in many places in the world, that implies that farms should have an individual mineral balance? Farms have to keep books of their mineral balances. As soon as a farmer has an output above a particular level, he is required to pay a levy. In The Netherlands this legislation is due within the next two years. Is it wise to do this on an individual farm basis, or would it be better to have this monitored at a higher level?

Bouma: The battle has to be won on the individual farms. When it is good business it will happen. But I would feel much more comfortable if that mineral balance were to be put in the context of dynamic properties in terms of nutrient fluxes of each particular soil type on the farm in its geological landscape context.

Groffman: It's most fair to do it at the farm or field scale, because then all farmers are being treated the same. However, if you measure the mineral balance at the catchment scale, the mineral balance of a farmer surrounded by forest could be different from that of another farmer surrounded by other farms. The question is: what are the analogies with other sources of pollution? In the USA we have air quality legislation that says that your factory can emit different levels of pollution depending on how clean the air is where you live. Whether or not we can do this in agriculture, I don't know. But if we want all farmers to be treated the same, then this sort of legislation has to be at the field scale and not at the catchment scale.

Rabbinge: The problem here is that in some areas it is much more difficult to stay below certain levels than in others, depending on the soil type. In this case, farmers are being treated differently by a common legislation.

We have been struggling with this legislation over the last year, and we decided that it would be based on the individual farm level. Although there are problems with this, we feel the individual farm-based mineral balance is the only way to proceed, and we have to accept that farmers of well endowed agricultural areas are conferred an advantage by this legislation.

Robert: In Minnesota, a split application of nitrogen during the growing season is already a best management practice (BMP) on sandy soils to limit risks of nitrogen leaching into groundwater. Determinations of N rates are based on quick on-site analysis of N concentration in potato petiole sap. Under pivot irrigation systems, split applications of N can be performed during water applications. New precision irrigation systems in development are capable of controlling application rates of each nozzle or groups of nozzles located on the pivot boom from precision application maps.

Rabbinge: Legislators can regulate either aims, or means. In my opinion it is best to dictate levels that must be met (e.g. less than 50 mg nitrate per litre of water) and then to allow individual farmers to decide how they are going to meet this requirement. This stimulates the entrepreneurial spirit of farmers and does not limit them to the strict, preconceived ideas of bureaucrats.

Groffman: That's a very important point. I agree that we should stipulate the standard and leave it up to the farmer to meet this however he chooses, in order to stimulate innovation. A analogous situation occurred in the US automobile industry. Some while ago the US government legislated that all automobiles must achieve a fuel efficiency of 27 miles per gallon, but they left it up to the manufacturers to decide how they would meet that standard. They achieved that standard very easily, although they complained quite a bit. Performance standards are generally better than design standards.

Bouma: Marc Voltz, I have a question that relates to your first point, concerning the heterogeneity in rooting systems. In modelling approaches it is common to look at the whole site under study and to try and derive a model for the entire site. However, in your vineyards there are obviously some vines that do very well and others that grow poorly. Couldn't you go in and look at vines that are doing well and measure, for example, the soil characteristics, root structures and moisture regimes of these to derive rules for managing the other vines, rather than modelling all of the vineyard?

Voltz: I agree in principle, but the problem with vineyards is that vinestocks grow for 40 years, after which they are usually replaced. This means that when the farmer manages his land he has to take into account what has been done previously, often by the former generation. This means that it's very difficult to do what you suggest with vineyards because both quantity and quality of grapes are not only affected by vineyard management in the year of harvest, but also by what has happened over the last 20 years or so. In vineyards, because of competition effects, there are both robust and weak vinestocks, and there is little you can do to improve the latter when the vinestocks are already several years old.

Burrough: Getting the basic management or sampling unit right is of great importance. This was illustrated by Philippe Lagacherie's work on basic soil units within this particular catchment as a model for mapping larger areas (Lagacherie et al 1995). Instead of looking at the basic sample as being something that's perhaps a kg soil or soil profile, he was looking actually at the little fields as hydrological units. This was this very successful because it removed a lot of the short-range variation that was not appropriate to the problem that he was looking at, which was the regional variation.

Mulla: Marc, during your talk you commented that you would be interested in knowing how to make better *in situ* measurements, in order to characterize variability. It has been apparent to me in some of my research that subsurface soil horizonation is critical in determining things like rooting patterns and subsurface moisture flow. In order to characterize this, have you ever thought of using non-invasive methods such as ground penetrating radar?

Voltz: I have thought about this, but I haven't had the opportunity to do it. In the vineyard we are studying there is a topsoil with a sandy sub-soil. The rooting depth is determined by the sedimentary layer below this which forms a barrier to the penetration of the roots. The depth of this layer is variable within the field and is hard to map. For this it would be useful to have non-intrusive methods such as ground penetrating radar.

Rabbinge: I have a technical question concerning rootstocks. In apple trees weak root stocks are desirable because they cause an increased harvest index. The consequence of this is that these apple trees are very sensitive to water shortages. Is this also the case for vines?

Voltz: I don't know exactly, because I'm not a vine physiologist, but I don't think so. In general, vinestocks are known to have very big rooting systems. Roots have been observed down to 30 m, and it's common to find them down to 10 m. We don't know why they grow this deep. In the situation we have studied, water extraction happens only in the first 2 m.

Rabbinge: For roots 30 m down the suction requirement for extracting water would need to be at least 20 bars.

Voltz: Yes, it's a problem.

Bregt: We are now talking about the precise measurement of soil variability, and we have discussed the relevance for environment impact, but if you're looking at crop production, especially when you're talking about vineyards with deep rooting systems, I get the feeling that the short-range spatial variability may be of minor importance for this type of crop. The crop is integrating or smoothing a lot of spatial variability in the soil. So you're measuring in detail, but you're not saying anything about the relationship between spatial variability and crop production.

Voltz: I agree that in the situation I presented the spatial variation in soil properties seems a less important factor than the rooting density. But the variation in soil properties is an indirect factor because it influences the rooting distribution. In this vineyard we also tried to investigate the soil moisture effects on the vinestocks and their yields. No clear relationship was found. In fact, because of the local heterogeneities and the fact that the extent of the rooting system of a given vinestock is unknown, it is very difficult to define at which scale soil water conditions should be measured. So we can't avoid studying the scales of heterogeneity in soil properties.

Burrough: It also means you have this problem of relating a point measurement of some index of plant performance.

McBratney: I guess what Arnold Bregt is saying is that each different kind of plant has some kind of integral scale, and it is much bigger for a vine than for a wheat plant.

Rabbinge: This depends on the conditions of the soil, the characteristics of the soil, the water table and other such factors. These factors dictate the form the rooting systems take. In the past, physiologists just looked above ground. Recently, we have had more detailed observations below ground, since we have developed rhizolabs that enable a more detailed monitoring of basic subsurface processes.

McBratney: This is going to be important because we have to address the question concerning which scale we are going to try and manage this variability at. This integral scale is obviously the minimum scale that you're going to work at, but it may not be the optimum one.

Burrough: That is extremely important also for linking remotely sensed imaging to crop performance, because we're dealing again with averaging over a particular area. Some of the work Arnold Bregt did many years ago on the Minderhoudhoeve farm demonstrated that variation in the 15 cm blocks is very different from that over 1.5 m. The difference was so extreme the patterns were not correlated with each other (Burrough 1991).

Meshalkina: I would like to add that the spatial distribution of the production may not correlate solely with the rooting systems. The distribution of different weed species in the field can be informative, too. Due to the strong competition between weed species on the field under different crops and weather conditions, the maps of the distributions may show the patterns of production variability.

Rabbinge: You're right. If you bring in weeds, diseases and pests, we know that the crops have to behave under different circumstances and still compete with weeds, of which there are a large variety. Weeds have a much more flexible response to changing conditions because there are many different weed species. The same is true of plant diseases: there are diseases of crops that are doing well and also of crops that are doing badly. In other words pests, diseases and weeds may have a flexible response that is not present in plants and crops.

I would like to summarize what we have been discussing this morning in nine points.

(1) We discussed the necessity of clarifying the level at which you are making your decisions: whether these are strategic, technical or operational decisions.

(2) We discussed at length spatial and temporal variation. Johan Bouma proposed that variation in time is often more important than spatial variation. It is the combination of both that is important in precision agriculture.

(3) The third point concerned looking forwards and backwards. A backward-looking approach is easier — in hindsight we can explain why things went wrong and what should be done better in the future. Looking forwards is much more complicated, more so because we are working with systems highly dependent on the weather, which we can only forecast accurately about three days in advance.

(4) The role of technology. In the past the role of technology was more reactive, but we would like to move into more proactive types of technologies, which creates a need for explanatory models in combination with summary models that are used for decision-making.

(5) How should we manage risk? It is possible to define critical levels which are acceptable in terms of management of the norms for the emission of nitrogen and phosphate, for example.

(6) We discussed why fine-tuning and precision agriculture are important. They may help to serve multiple goals, not only the goal of productivity, but also the goal of trying to eliminate negative side-effects.

(7) The next point, which became clear in the last discussion we had, concerned where we should measure pollution. Should we monitor it on a watershed level, a catchment level or on a field level? How can we scale up and scale down from one level to another?

(8) We also talked about norm setting. Does this involve dictating means, or merely setting the aims? We agreed that setting aims is much better than regulations that dictate means.

(9) Finally, we discussed the rooting depths and rooting systems, knowledge of which are vital for us to develop precision agriculture further. These may differ between annual and perennial crops and also with soil characteristics and water tables.

References

Burrough PA 1991 Sampling designs for quantifying map unit composition. In: Wilding L, Mausbach M (eds) Spatial variabilities of soils and landforms. Soil Science Society of America Special Publication 28, Madison, WI, p 89–125

Lagacherie P, Legros JP, Burrough PA 1995 A soil survey procedure using the knowledge of soil patterns established on a previously mapped reference area. Geoderma 65:283–301

Modelling non-stationary spatial covariance structure from space–time monitoring data

Pascal Monestiez*, Wendy Meiring†, Paul D. Sampson‡ and Peter Guttorp‡

**Unité de Biométrie, INRA, Domaine Saint Paul, Site Agroparc, 84914 Avignon Cedex 9, France, †Climate and global dynamics, NCAR, Boulder, CO 80307, USA and ‡Department of Statistics, University of Washington, Seattle, WA 98195, USA*

Abstract. Accurate interpolation of soil and climate variables at fine spatial scales is necessary for precise field management. Interpolation is needed to produce the input variables necessary for crop modelling. It is also important when deciding on regulations to limit environmental impacts from processes such as nitrate leaching. Non-stationarity may arise due to many factors, including differences in soil type, or heterogeneity in chemical concentrations. Many geostatistical methods make stationarity assumptions. Substantial improvements in interpolation or in the estimation of standard errors may be obtained by using non-stationary models of spatial covariances. This paper presents recent methodological developments for an approach to modelling non-stationary spatial covariance structure through deformations of the geographic coordinate system. This approach was first introduced by Sampson & Guttorp, although the estimation approach is updated in more recent papers. They compute a deformation of the geographic plane so that the spatial covariance structure can be considered stationary in terms of a new spatial coordinate system. This provides a non-stationary model for the spatial covariances between sampled locations and prediction locations. In this paper, we present a cross-validation procedure to avoid over-fitting of the sample dispersions. Results concerning the variability of the spatial covariance estimates are also presented. An example of the modelling of the spatial correlation field of rainfall at small regional scale is presented. Other directions in methodological development, including modelling temporally varying spatial correlation, and approaches to model temporal and spatial correlation are mentioned. Future directions for methodological development are indicated, including the modelling of multivariate processes and the use of external spatially dense covariables. Such covariates are frequently available in precision agriculture.

1997 Precision agriculture; spatial and temporal variability of environmental quality. Wiley, Chichester (Ciba Foundation Symposium 210) p 38–51

Fine spatial interpolation of soil variables at the field scale, and of climate variables at the small region or the farm scale, are essential for precision agriculture. This is to produce the input data necessary for crop models and for field management at very local scale (Robert 1993). Obtaining such precision in spatially accurate farming also provides a way to study and decide on controls to limit environmental pollution, such as that caused by nitrate leaching (Finke 1993).

Although geostatistical techniques have been used for many years in soil science and climate interpolation, there is still a strong need for improving interpolation methods and precision. Improvements in fine spatial interpolations are possible for at least two reasons. Firstly, multivariate statistical approaches to spatial modelling may use the vast amount of external data which are now available from remote sensing (aerial photography), soil databases, local topography and the exhaustive yield map from the year before. Papers presenting such multivariate model improvements and their use in model comparison include Odeh et al (1995) and Knotters et al (1995). Secondly, geostatistical approaches have been developed in the earth sciences where there often are data from only a single realization of the studied phenomena. Most techniques are not well adapted to precision farming or environmental process studies where the spatial phenomena concerning a given field will be studied for several years. When a process is observed over time, the data can be used as pseudo-replications or in a spatiotemporal model. Basic hypotheses of geostatistics, including stationarity or ergodicity, are no longer necessary operational hypotheses. In many cases, spatial non-stationarity is more realistic for modelling the real world, and we can now go further in improving spatial models. This will lead to improvements of local interpolation precision.

In this paper, we address primarily the problem of modelling and estimating the non-stationary spatial covariance structure in levels of a field process at arbitrary locations (both monitored and unmonitored), based on records from N point monitoring sites. The fundamental idea of the non-stationary spatial covariance approach is to compute a deformation of the geographic plane so that the spatial covariance structure can be considered stationary (and isotropic) in terms of a new spatial coordinate system. In recent years the methodology has been developed by a number of authors and applied in analyses of solar radiation (Sampson & Guttorp 1992), acid precipitation (Guttorp et al 1993, Guttorp & Sampson 1994, Mardia & Goodall 1993), rainfall (Monestiez et al 1993, Meiring 1995), air pollution (Brown et al 1994) and tropospheric ozone (Guttorp et al 1994, Meiring 1995). Potential applications in agriculture include the assessment of underground water quality in agricultural districts, and the accurate spatial prediction of air temperature (which is an input to a growth or yield model) when topography or local context induce non-stationarity. Although the example presented here is not at the field scale, the methodology for fine interpolation of rainfall at local scale may be extended to the field scale for irrigation nonuniformity (Or & Hanks 1992) and for soil water monitoring.

Model and model estimation

The environmental process is measured at N point monitoring sites $x_1, \ldots, x_N$, which recorded observations at each of T time points. We denote these records by $Z(x_i, t)$, $i \in \{1, \ldots, N\}$, $t \in \{1, \ldots, T\}$; and consider them to be a sample from the space–time process $Z(x, t)$. For most of this paper we assume independence in time, although we discuss some extensions for modelling spatial and temporal correlation later on.

We model the *spatial dispersion* function, defined as $D(x, y) = \mathrm{Var}\,(Z(x, t) - Z(y, t))$ for each pair of geographic locations x and y, by

$$D(x, y) = \gamma_\theta (\| f(x) - f(y) \|), \tag{1}$$

where $f(\cdot)$ represents a smooth (bijective) transformation of the geographic coordinate system, and γ_θ represents an isotropic variogram function with parameters θ.

We denote the sample estimate of the dispersion between the ith and jth monitoring sites by d_{ij}; and the fitted dispersion between the ith and jth monitoring sites by $\widehat{d_{ij}} = \gamma_{\hat{\theta}} (\| \hat{f}(x_i) - \hat{f}(x_j) \|)$. For analysis and modelling we usually standardize the site-specific time series to variance 1, in which case we can compute $d_{ij} = 2(1 - r_{ij})$, where r_{ij} is the empirical correlation between the observations at sites x_i and x_j. We refer to the geographic coordinate system as the G-plane, and the transformed coordinate representation as the D-plane.

We calculate the G-plane to D-plane transformation using a pair of thin-plate splines. The first spline maps the two-dimensional G-plane into the first coordinate of the D-plane representation and the second spline maps the G-plane coordinates into the second coordinate of the D-plane representation. The splines are fully determined by the D-plane locations of the monitoring sites, so the model-fitting involves estimation of these N D-plane locations together with the parameters of the isotropic variogram model γ_θ. We minimize the objective criterion

$$C_{\theta, f, \lambda} = \sum_{j=2}^{N} \sum_{i=1}^{j-1} \left[\frac{d_{ij} - \widehat{d_{ij}}}{\widehat{d_{ij}}} \right]^2 + \lambda\,\mathrm{BEP} \tag{2}$$

with respect to θ and the D-plane coordinates, where BEP denotes a *bending energy penalty* for the transformation. Two of the N monitoring site locations are held at their geographic locations in order to fix the location, scale and orientation of the D-plane representation. The weighted least squares term accounts for the heterogeneity in variance of the sample dispersions d_{ij} (see Cressie 1985), while the bending energy penalty controls the smoothness of the computed deformation. The so-called bending energy of an $\mathbb{R}^2$ to $\mathbb{R}$ thin-plate spline mapping g, is

$$\int_{\mathbb{R}^2} \left[\left(\frac{\partial^2 g}{\partial x^2} \right)^2 + 2 \left(\frac{\partial^2 g}{\partial x \partial y} \right)^2 + \left(\frac{\partial^2 g}{\partial y^2} \right)^2 \right] dx\, dy \tag{3}$$

For each of the thin-plate spline mappings, this integral can be expressed simply as a quadratic form in the D-plane coordinates. The total BEP of equation (2) is the sum of the bending energies for the two thin-plate spline mappings, a sum which is invariant under translation and rotation of the coordinate systems.

The choice of smoothing parameter is extremely important. For low values of λ we may 'over-fit' the sample dispersions (in view of their sampling variability) and find that the G-plane to D-plane mapping 'folds' (is not bijective), something which is generally uninterpretable. For high values of λ, the G-plane to D-plane mapping approaches an affine transformation, corresponding to a stationary or homogeneous

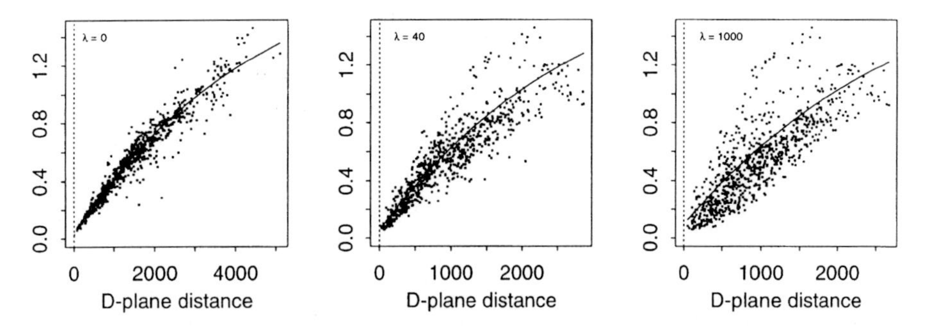

FIG. 1. Fitted exponential variograms with nugget, as a function of D-plane distance, for three values of λ. The spatial dispersions of the vertical axis correspond to spatial correlations ranging from 1 (for $d_{ij} = 0$) to about 0.3 (for $d_{ij} = 1.4$).

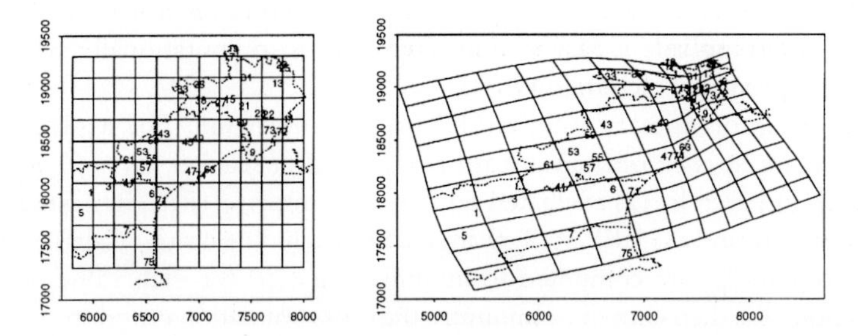

FIG. 2. (*Left panel*) Relative locations of the 39 sites used in the estimation. Site numbers are those used in Monestiez et al. (1993). (*Right panel*) Deformation map for $\lambda = 40$. The boundaries of three French departments are drawn in each panel.

model with elliptical anisotropy. Figure 1 shows fitted dispersion models for time series of altitude-adjusted 10 day aggregated precipitation records from November and December of 1975 through 1992 (108 records per site) for 39 sites from the Languedoc-Roussillon region of southern France. Results for three values of λ are shown, using an exponential variogram model with nugget. Figure 2 shows the corresponding G-plane to D-plane deformation map for $\lambda = 40$. Sites in the northeast region are relatively compressed in the D-plane representation because of relatively higher spatial correlation there.

The next section addresses the choice of λ on simulated data, and then the section after that presents a bootstrap approach for assessment of variability of the resulting covariance estimates using the rainfall data set.

Choosing the smoothing parameter

In this section we address the choice of λ using a simulation experiment. We compare two different approaches for the specification of λ, along with the extreme cases of $\lambda = 0$ (no smoothing) and $\lambda = \infty$ (yielding an affine G-plane to D-plane mapping, corresponding to a stationary, elliptically anisotropic model).

The first of the two procedures uses a simple stopping rule for a series of decreasing values of λ. The first term of equation (2) decreases monotonically as λ decreases, reaching a minimum value at $\lambda = 0$, corresponding to the fitting of a model without any smoothing constraint. This case generally results in overfitting of the deformation with values of the weighted least-squares component of the criterion less than the expected value under the true underlying model. In the case of Gaussian data where the underlying model (θ and $f(\cdot)$) is known (and without standardizing the time series to variance 1), the d_{ij} are simply empirical variances so that the expected value of the first term of the criterion is:

$$E\left[\sum_{i \neq j \in \mathcal{N}} \left(\frac{d_{ij} - \gamma_\theta(f(x_i) - f(x_j))}{\gamma_\theta(f(x_i) - f(x_j))}\right)^2\right] = N(N-1)/T \tag{4}$$

where $\mathcal{N} = \{1, \ldots, N\}$. In order to avoid overfitting we choose that value of λ in a decreasing sequence, for which the first term of the criterion reaches this expected value. This stopping rule is easy to implement and is computationally simpler than our second approach, which follows.

The second procedure is a cross-validation minimization. For a given λ, each observation site x_i is eliminated in turn from the data set. Let $C_{\theta,f,\lambda}(i)$ denote the criterion $C_{\theta,f,\lambda}$ given by (2) evaluated on $N-1$ monitoring sites after excluding site x_i. (Both the sum and the BEP in (2) change when excluding a site.) A deformation $\hat{f}_{i\lambda}$ and a variogram $\gamma_{\hat{\theta}_{i\lambda}}$ are computed to minimize $C_{\theta,f,\lambda}(i)$ for each value of λ under consideration. λ is then chosen to minimize the cross-validation criterion

$$C_{cv\lambda} = \sum_{i \in N} \sum_{j \in N - \{i\}} \left(\frac{d_{ij} - \gamma_{\hat{\theta}_{i\lambda}}(\hat{f}_{i\lambda}(x_i) - \hat{f}_{i\lambda}(x_j))}{\gamma_{\hat{\theta}_{i\lambda}}(\hat{f}_{i\lambda}(x_i) - \hat{f}_{i\lambda}(x_j))}\right)^2. \tag{5}$$

Simulation experiment

The function f used in this simulation experiment was computed using a pair thin-plate splines, as described in the section on the model. Nine geographic locations and their D-plane locations were used to define a case-study transformation which was clearly non-linear, but smooth (Fig. 3). A grid of $N = 36$ points is taken to define the observation sites in the G-plane and γ_θ was specified as an isotropic stationary exponential variogram model with no nugget.

Using the computed image of the $N = 36$ sampled sites in the D-plane together with the variogram model, we computed the true underlying covariance matrix according to the model of equation (1). We then simulated T independent sets of 36 multivariate normal variables. Simulations were run for time series of length $T = 25$, 100 and 400.

Twenty simulations were run for each of the three values of T. For each simulation, f and γ_θ were estimated using each of 27 different values of λ between 0 and 10 000. An example is given in Fig. 4 for one simulation of length $T = 25$.

Comparing the approaches for choosing λ

We used a larger set of validation sites in order to compare the different choices of λ. The test grid was defined as a 9×9 grid of points with corner points (0,0), (1,0), (1,1) and

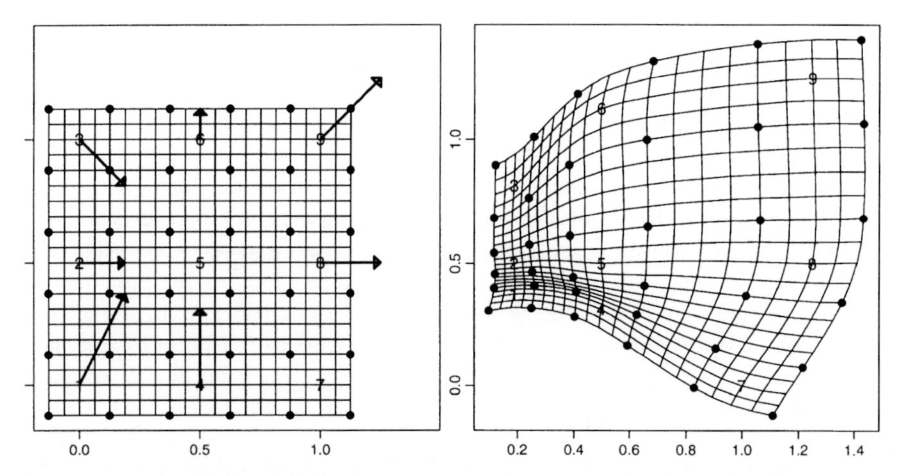

FIG. 3. (*Left panel*) The geographic space. Numbers indicate the nine points and their translations used to define f; the 36 observation sites are marked by dots. (*Right panel*) The transformed D-space in which the process is isotropic.

(0,1). (These grid points are at the intersections of every other horizontal and vertical grid line in Fig. 3.) The D-plane locations of the 81 points were computed using the estimated f for each value of λ, and the dispersions were then estimated for all pairs of combinations of the 81 sites. The validation criterion is the Mean Square Error comparing the predicted and the theoretical dispersions computed using the theoretical f and variogram model from which we simulated. The mean square is computed over all 3240 pairwise dispersions separately for each of the 20 simulations. For $T = 25$, estimated values of the pairwise dispersions are plotted against the theoretical values in Fig. 5 for three values of λ: zero, the value minimizing the cross-validation criterion, and the value resulting from the simple stopping rule.

Each of these three values of λ resulted in better predictions of the true dispersions (i.e. a lower validation criterion) than a model with a homogeneous affine

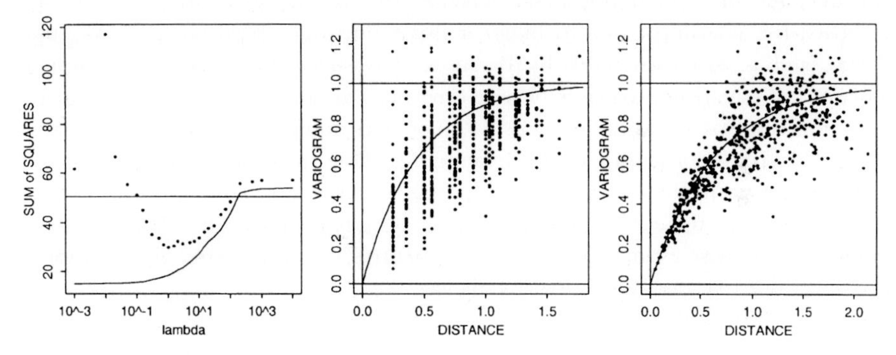

FIG. 4. (*Left*) Criterion versus λ: solid line gives the minimization criterion without the bending energy term, dots give the cross-validation criterion. (*Middle*) Variogram plot for $\lambda = \infty$ (homogeneous affine deformation). (*Right*) Variogram plot for $\lambda = 1$ close to the minimum of the cross-validation criterion.

transformation ($\lambda = \infty$). The fits with no smoothing ($\lambda = 0$) resulted in deformations f that folded (as demonstrated in the lower left panel of Fig. 5). None the less, dispersions are reasonably well predicted at short geographic distances, which are not much affected by the folding. Much larger cross-validation errors may be introduced for sites further apart in the D-plane.

The simple stopping rule generally stopped too early in the decreasing sequence, resulting in too much smoothing and a corresponding bias in the estimates, especially for short spatial lags. The shortest D-plane distances were, relatively speaking, not as small as they should have been according to the true model (compare the right panel of Fig. 3 and the lower right panel of Fig. 5). This oversmoothing did not affect the accuracy of the predictions of dispersions at longer range, but these are less important in kriging calculations. The value of λ determined by cross-validation resulted in more accurate predictions.

Comparisons of the results of the cross-validation criterion (Mean Squared Error) across the 20 simulations are summarized in Fig. 6 for each of the values of T. The main result is that the determination of λ by cross-validation yields nearly the optimal value (as determined by minimum of the validation Mean Squared Error). The simple stopping rule derived from the approximate expected value of the weighted least-squares criterion performed poorly, even for the longest time series ($T = 400$). As expected, for sufficiently large T, no smoothing is necessary, but for short time series, the computationally intensive cross-validation calculation appears necessary.

Assessment of variability

When assessing variability of kriging estimates, researchers most often assume a known covariance structure; i.e. they ignore the uncertainty in the estimated covariance structure. (Some Bayesian approaches to spatial estimation provide an exception: see, for example, Brown et al 1994.) For the approach described in this paper, it is important to ask whether local structure in the covariance field may be due to a few influential points, or whether this is true structure. A bootstrap approach for assessment of variability of this estimator was proposed in Meiring (1995), and is based on work by Beran & Srivastava (1985) on bootstrapping functions of covariance matrices. Whole time slices are resampled with replacement from the original data, to generate M bootstrap samples each composed of T resampled vectors of length N. By resampling whole time slices, we maintain the spatial structure in the original data. The spatial correlation structure is estimated based on each of these M bootstrap samples. Kriging estimates could be obtained using each of these estimated covariance structures, to assess better the additional variability due to estimation of the covariance.

Figure 7 shows sample from the rainfall data set, fitted (for $\lambda = 40$), and bootstrapped dispersions between site 45 and the other 38 sites used in this study. The boxplots of the bootstrapped dispersions are based on 50 bootstrap samples from the original data. The spatial covariance modelling approach smoothes the sample dispersions, which are not always covered by the bootstrap distributions.

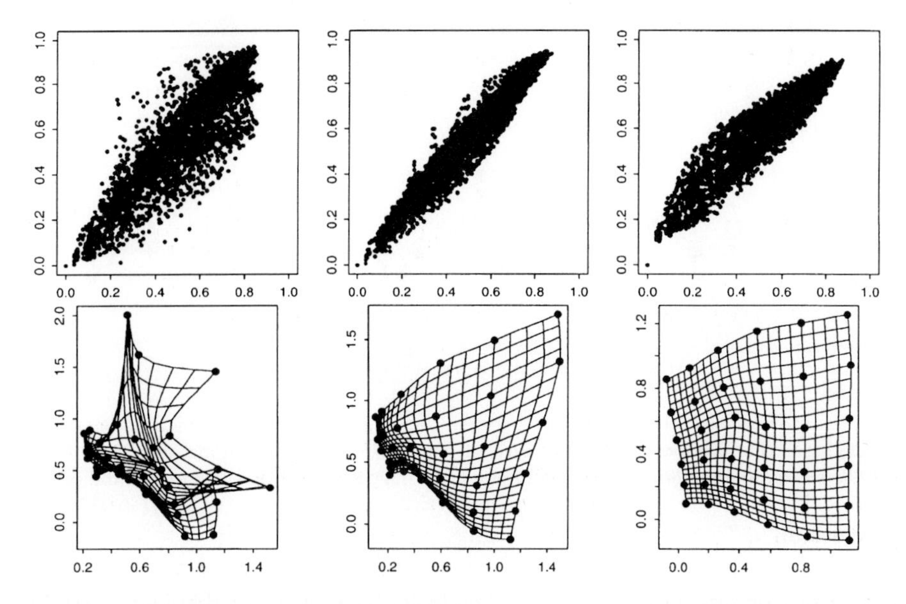

FIG. 5. (*Upper row, left to right*) Predicted variograms for all pairs of the test grid versus theoretical values (along the x-axis) (i) without smoothing ($\lambda = 0$) (ii) for λ minimizing the cross-validation criterion (iii) for λ resulting from the expected value of the criterion. (*Lower row*) Corresponding estimated deformation f for the same three choices of λ.

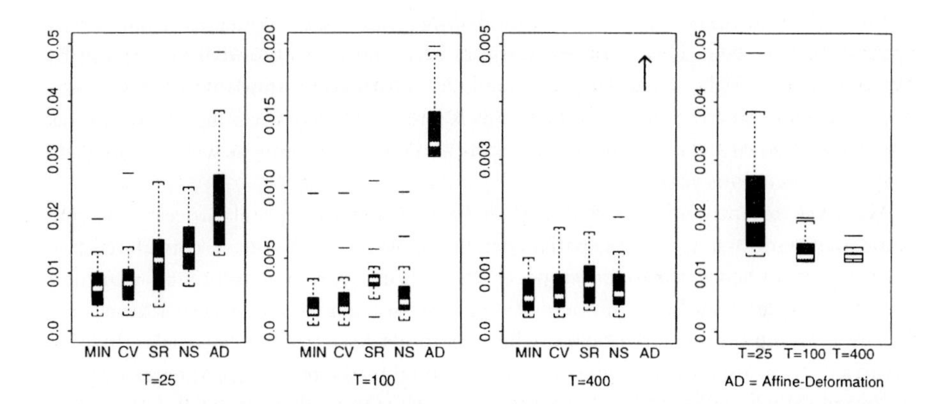

FIG. 6. Box plots of the Mean Squared Error (MSE) validation criterion computed for the 9×9 test grid for each of 20 simulations for T values of 25, 100 and 400. MIN, λ yielding the absolute minimum MSE; CV, λ chosen by the cross-validation criterion; SR, λ chosen by the simple stopping rule; NS, $\lambda = 0$ (no smoothing/bending energy penalty); AD, $\lambda = \infty$ (infinite smoothing/homogeneous affine model). Note that the vertical axes differ from plot to plot due to the greater accuracy of the fitted values for large T. The fourth plot on the right shows, on a common scale, the MSE criteria for the AD case.

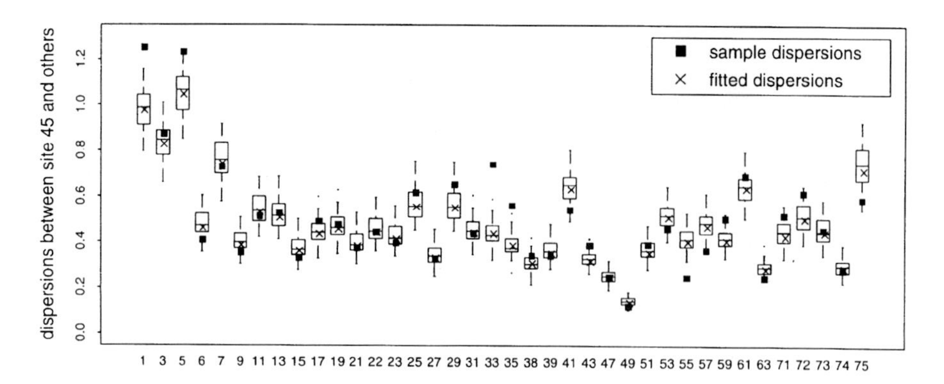

FIG. 7. Boxplots of fitted dispersions between site 45 and sites indicated by horizontal axis labels, based on 50 bootstrap samples each of length 108 from the original data, and using $\lambda = 40$. The sample and fitted (shown in Figure 1, $\lambda = 40$) pairwise dispersions are also indicated.

Discussion

Our spatial deformation model seems heuristically justified for the many areas of application that we have considered and referenced. However, both of the components of the model — the deformation mapping and the variogram function — need not be restricted to the classes of functions that we currently use. The thin-plate spline was chosen for convenience. Smoothing using the bending energy penalty in order to eliminate folding also results in mappings that may be smoother than necessary at small spatial scales. A different class of mappings, or equivalently, a different approach to smoothing or regularization may prove useful.

One obvious extension of this modelling framework concerns multivariate spatiotemporal data. This has been addressed by Mardia & Goodall (1993) and by Brown et al (1994). However, the explicit deformation modelling of a cross-variogram has yet to be developed and this is a real challenge for integrating external data coming from soil databases, topography and remote sensing, as well as from yield maps from previous years.

We have recently addressed the modelling of spatiotemporal processes having important temporal as well as spatial correlation structure. We have considered two different approaches to spatiotemporal estimation. The first approach considers spatially varying time series models. We first pre-whiten these individual time series before applying the deformation modelling methods discussed here. Spatial estimation is carried out using the pre-whitened residuals and then the resulting estimates are post-coloured (Meiring 1995). An alternative approach is to apply the spatial deformation modelling method to the space–time covariance structure using the bilinear model of co-regionalization as suggested by Rouhani & Wackernagel (1990). The temporal aspect cannot be ignored in the future for precision agriculture applications. It may result in a substantial gain to use the previous years observations in a spatiotemporal model, as compared with considering the data from these years as external variables. The studied phenomena may also be non-stationary in time.

Currently, the fitting of our models is a challenging numerical problem with dimensionality roughly proportional to the number of fixed monitoring sites. This makes the cross-validation and bootstrapping procedures presented here extremely demanding computationally. The dimensionality of the problem leads us to ask if this deformation model might be fitted using Markov Chain Monte Carlo techniques, which have proved so useful in similarly high-dimensional problems of image analysis. Some aspects of the problem would be similar to image analysis, particularly if the model integrates remote sensing images as covariates (Robert 1993).

Acknowledgements

Funding from the Electric Power Research Institute, INRA, the National Research Center for Statistics and the Environment, and the Geophysical Statistics Project at the National Center for Atmospheric Research is gratefully acknowledged. The National Center for Atmospheric Research is sponsored by the National Science Foundation. Although the research described in this article has been funded in part by the United States Environmental Protection Agency through agreement CR825173-01-0 to the University of Washington, it has as not been subjected to the Agency's required peer and policy review and therefore does not necessarily reflect the views of the Agency and no official endorsement should be inferred. Figures and parts of the text are reproduced from Meiring et al (1997) with kind permission of Kluwer Academic Publishers.

References

Beran R, Srivastava MS 1985 Bootstrap tests and confidence regions for functions of a covariance matrix. Ann Stat 13:95–115 (Correction 1987 Ann Stat 15:470–471)

Brown PJ, Le ND, Zidek JV 1994 Multivariate spatial interpolation and exposure to air pollutants. Can J Stat 22:489–505

Cressie NAC 1985 Fitting variogram models by weighted least squares. Math Geol 17:563–586

Finke PA 1993 Field scale variability of soil structure and its impact on crop growth and nitrate leaching in the analysis of fertilizing scenarios. Geoderma 60:89–107

Guttorp P, Sampson PD 1994 Methods for estimating heterogeneous spatial covariance functions with environmental applications. In: Patil GP, Rao CR (eds) Handbook of statistics XII: environmental statistics. Elsevier Science Publishers, New York, p 663–690

Guttorp P, Le ND, Sampson PD, Zidek JV 1993 Using entropy in the redesign of an environmental monitoring network. In: Patil GP, Rao CR (eds) Multivariate environmental statistics. Elsevier Science Publishers, Amsterdam, p 175–202

Guttorp P, Meiring W, Sampson PD 1994 A space–time analysis of ground-level ozone data. Environmetrics 5:241–254

Knotters M, Brus DJ, Oude Voshaar JH 1995 A comparison of kriging, co-kriging and kriging combined with regression for spatial interpolation of horizon depth with censored observations. Geoderma 67:227–246

Mardia KV, Goodall CR 1993 Spatial–temporal analysis of multivariate environmental monitoring data. In: Patil GP, Rao CR (eds) Multivariate environmental statistics. Elsevier Science Publishers, Amsterdam, p 347–386

Meiring W 1995 Estimation of heterogeneous space–time covariance. PhD Dissertation, University of Washington, Seattle, WA, USA

Meiring W, Monestiez P, Sampson PD, Guttorp P 1997 Developments in the modelling of nonstationary spatial covariance structure from space–time monitoring data. In: Baafi EY, Schofield NA (eds) Geostatistics Wollongong '96. Kluwer Academic Publishers, Dordrecht, p 162–173

Monestiez P, Sampson PD, Guttorp P 1993 Modelling of heterogeneous spatial correlation structure by spatial deformation. Cahiers de geostatistique, Fascicule 3, Compte Rendu des Journees de Geostatistique, 25–26 May 1993, Fontainebleau. Published by the École Nationale Supérieure des Mines de Paris

Odeh IOA, McBratney AB, Chittleborough DJ 1995 Further results on prediction of soil properties from terrain attributes: heterotopic cokriging and regression-kriging. Geoderma 67:215–226

Or D, Hanks J 1992 Soil water and crop yield spatial variability induced by irrigation nonuniformity. Soil Sci Soc Am J 56:226–233

Robert P 1993 Characterization of soil conditions at the field level for soil-specific management. Geoderma 60:57–72

Rouhani S, Wackernagel H 1990 Multivariate geostatistical approach to space–time data analysis. Water Resources Res 26:585–591

Sampson PD, Guttorp P 1992 Nonparametric estimation of nonstationary spatial covariance structure. J Am Stat Assoc 87:108–119

DISCUSSION

Rasch: Did I understand correctly that the function depends on the complexity of the mapping, f? What is the number of parameters of this function? Does it play a role?

Monestiez: The spline can be seen as non-parametric transformation. When the number of measured sites increases, so does complexity. Relative to the minimization, we can run into major difficulties if this number reaches large values.

Rasch: If this is the case I am surprised that you recommend the jackknife procedure. In our simulation experiments where we selected between 10 models, when all the criteria were used in combination with the corresponding jackknife criteria we found that the original criteria did better than the jackknife criteria (Otten et al 1996).

Monestiez: I'm not sure I understand your point. It's true that if the deformation belongs to a class of parametric functions then, as you suggest, we can discuss the choice of the jackknife criterion. Here we preferred initially to stay with a non-parametric function. In our context the jackknife criterion relies only on the penalty parameter and not on the whole set of parameters. We are now looking for a new family of deformations, and we would prefer the parametric even if it is with a lot of parameters, but until now this has not been an option.

Rabbinge: Dieter Rasch, in which sorts of procedures do you use non-parametric methods as opposed to parametric methods?

Rasch: The problem of whether or not to use jackknife is independent of whether we use parametric or non-parametric approaches. We used parametric methods in non-linear regression, and especially in growth research, and we selected out of 10 functions the best one, following different criteria. The first step in our work was to evaluate the criteria to find which was best for selecting between functions. It was a two-step selection: first we selected good criteria by using simulation methods and then we used these criteria to select one of the corresponding models. In the first step, selecting the criteria, we found that the original criteria were not improved by the jackknife versions.

Rabbinge: But if you are developing your criteria in that way, isn't there a danger of losing independence, because you are working with an iterative procedure such that you are improving the quality of the final results by coupling them to the input data? That will increase the danger of losing the independence of the criteria that you used.

Rasch: We had 10 functions, we simulated data using these functions and then we used the criteria to find out which function fits the best. This resulted in a 10×10 matrix, which we used to select the best criterion. This was a simulation experiment using no data solely for the purpose of finding out which criterion works best. Later on we used only this criterion. There is no dependence because each data real set is independent of our simulated data for evaluating criteria.

Stein: Pascal Monestiez, in the final point in your presentation, there was a combination of processes. I think it was the spatiotemporal data with an origin somewhere in time. The process then expands and becomes visible in space. You don't have the alternative, i.e. a process in space which becomes visible in time. Have you ever considered analysing spatial data for finding the origin in time? And did you ever combine the data with an auto regressive or moving average process?

If you have a time series and you try to model this with an auto regressive process, it might be successful to combine this auto regressive process with the spatial process.

Monestiez: In fact we suppose orthogonality, or in other words 'independence' between time and space. If you do not suppose this, you will have to compute covariance function between time and space, so you go back to the single realization context. It may be difficult to fit the whole model in space and time from a single realization of the process. So, in a sort of compromise, we go further but not completely with crossing time and space.

Stein: You showed how to average realizations of a process in time and you combined it to get a better estimate of the unknown variable. Perhaps you can do more by adding the process-based information to the stochastic model that you applied.

Monestiez: If you have some strong structure in time, you will in fact be doing that on the residual: you will model the time process for spatial interpolation of the residuals. But you need orthogonality between the modelled term and the residual.

Thompson: I like this approach of using the different times as replications, but whenever I have looked at weather-related data, they show this long term correlation over months, years and decades. It seems to be the same problem that we work with in space; we have a large set of data but just one realization of the process. Statistical inference in that case is usually done by modelling covariance structure, and it may well be that we have to consider that approach also over time.

Monestiez: That is true. In the example I showed, we deal with weather. To get temporal replications we selected, for example, the month of December over 20 years. So we did not mix summer data with winter data. It's very difficult to say that there is a correlation between what happened in December one year and what happened in December the following year.

Gómez-Hernández: The main limitation of your method is basically that it must assume some kind of time independence. Your simulation experiment was with time-independent data. Have you tried to do the same thing with time-dependent data to try to prove that the method is robust in that situation?

Monestiez: No, we haven't. The result you get, for example, with 10 independent days, will be the same if you create dependence with 28 or 30 days, so you have to do more replications in time if you have dependence. What is important is that the dependence between two dates decreased with the time lag, so if observations are far enough apart in time, you can consider them to be independent replications.

Gómez-Hernández: If there is a strong time dependence, you would tend to overestimate the variances if you assume time-independent data.

Monestiez: If there is a strong time dependence, you need a robust and well fitted model of time series before fitting the spatial process. If you have a trend, you fit it and work on residuals.

Gómez-Hernández: It is not a trend: you can still have time correlation without the presence of a trend. But in that case, as you say, you are back to square one, the single realization problem.

Monestiez: If you get a strong correlation in time, you will get poor estimates of spatial covariance as a consequence of the error on the spatial deformation. Poor estimates don't mean systematic over-estimation of the variances.

Burrough: Exponents of thin-plate splines interpolation, which you mentioned, have been using thin-plate splines in a multivariate way to map rainfall, temperature and all sorts of things. These are easy-to-use techniques which can be incorporated into GIS. Your technique requires a considerable amount of expertise to drive it, at a level ordinary users of GIS would not necessarily have. What's the practical advantage of your technique compared with a relatively quick and dirty technique using the multivariate spline interpolation? How much more juice do you get out of the data?

Monestiez: We tried to quantify what we gained with this method on an actual example with a test validation data set. If you look at the improvement on a global point of view, for example a spatial average of the mean-square error of the interpolation, you may be disappointed by the gain in precision, which in our case was around 5%.

In fact, this average hides interesting local gains in precision. In many examples, there are large areas where ordinary kriging works well and where all the methods will be equivalent, and some quite different zones where we are far from the stationarity and where our method can locally improve a lot of the kriging interpolation. If these zones are just as important to you as the rest of the field, our method may locally substantially improve the interpolation.

Burrough: But can't you also adapt to your local non-stationarity by taking different domains and computing different variograms for those (i.e. use stratified or universal kriging)?

Monestiez: But then you are faced with the problem of variogram estimation if it is too local. Your solution may be right if the sampling is dense enough locally.

Rabbinge: I think I need to make it clear that 'simulation' is being used here in a different way to the normal production ecological use of the term. Here, simulation is being used to generate outcomes on the basis of formulae which are of a purely mathematical nature, and the resulting outcomes are then used to test against variable criteria. So simulation here is a tool in a mathematical procedure, whereas simulation in production ecological terms involves developing a model of the processes which occur (be it on a physical, pedological or chemical nature), integrating that information in a simulation model and comparing the behaviour of that model with reality, in order to get better insight into how these processes interact.

McBratney: When you get into your combine harvester and generate all these yield data, you have many thousands of data points in each field. Using these you can compute local variograms within a moving window easily and you see clearly that this covariance structure is not stationary: it changes from place to place. Because of the vast amount of data you can do it that way: you don't have to go through these fancy transformations. I think the point of your technique is that this transformation from the G-plane to the D-plane is useful. The D-plane is the more interesting one to be working in — this is the one in which the precision farmer has to optimize, for example, his sampling. What sort of decisions would it be useful to make in the D-plane? As I'm in Australia, I'm always worried about water, so I'm going to set up a monitoring network and put in some capacitance probes to measure soil moisture on a grid in the D-plane and then I'm going to transform that back to the geographic space. They're not going to be on a grid, they're going to be irregularly spaced, but that's exactly what your technique would be good for. It is a brilliant optical solution to sampling problems.

Stein: These techniques can be very useful if you have sparse spatial data and some more replicates in time. An example could be on say 10–20 observation points, and more than 10 temporal replicates, such as in some monitoring networks.

McBratney: Is the transformation from the D-plane back to the geographic one unique? In other words, can you go from G to D and back again to G, and end up in the same place?

Monestiez: We haven't found a mathematical form that is easily reversible. If we find one it should be easier to handle. So we arbitrarily defined the transformation from one space to another. Reverse function is numerically computed if necessary. Regarding the uniqueness of the estimation, you can prove the uniqueness of the estimation in an asymptotic context if the variogram is strictly monotonous.

Rabbinge: This technique has been used in radiation, acid rain, rainfall and also in air pollution. I would like to see how you can use that type of technique not only on a regional level but also on an individual field level. When you work on a regional level, very often data are lacking. For instance, you might have sparse rainfall data. For example, in Australia in the recent past there were only two places where there was continuous measurement of radiation. Can you use this type of technique to overcome such a lack of sampling data?

Monestiez: No, the space transformation would not be identifiable with only two sampling locations.

Reference

Otten A, Rasch D, van Wijk H 1996 Evaluation of criteria for the selection of models in non-linear regression. In: Faulbaum F, Bandilla W (eds) SoftStat '95: advances in statistical software. Lucius et Lucius, Stuttgart, p 489–496

Ecological constraints on the ability of precision agriculture to improve the environmental performance of agricultural production systems

Peter M. Groffman

Institute of Ecosystem Studies, Box AB, Millbrook, NY 12545, USA

Abstract. In this paper, I address three topics relevant to the ability of precision agriculture to improve the environmental performance of agricultural production systems. First, I describe the fundamental ecological factors that influence the environmental performance of these systems and address how precision agriculture practices can or cannot interact with these factors. Second, I review the magnitude of the ecological processes that we hope to manage with precision agriculture relative to agricultural inputs to determine whether managing these processes can significantly affect system environmental performance. Finally, I address scale incongruencies between ecological processes and precision agriculture techniques that could limit the ability of these techniques to manage variability in these processes. The analysis suggests that there are significant ecological constraints on the ability of precision agriculture techniques to improve the environmental performance of agricultural production systems. The primary constraint is that these techniques do not address many of the key factors that cause poor environmental performance in these systems. Further, the magnitude of the ecological processes that we hope to manage with precision agriculture are quite small relative to agricultural inputs and, finally, these processes vary on scales that are incongruent with precision management techniques.

1997 Precision agriculture: spatial and temporal variability of environmental quality. Wiley, Chichester (Ciba Foundation Symposium 210) p 52–67

This paper addresses several aspects of how ecological processes affect the ability of precision agriculture to improve the environmental performance of agricultural production systems. Much of the evaluation will be done by comparing processes in natural ecosystems with those in agricultural production systems. Such a comparison is useful because natural ecosystems are under strong pressure to minimize nutrient and soil losses (otherwise they would not persist). Many studies have examined structural and functional aspects of natural ecosystems that contribute to their high environmental performance and long-term sustained production. It is useful to consider how these aspects are expressed in agricultural production systems and how they are affected by precision agriculture techniques. The specific topics that will be addressed are:

(1) What are the fundamental ecological factors that influence the environmental performance of agricultural production systems? How do precision agriculture practices interact with these factors?
(2) What is the magnitude of the variation in ecological processes that we hope to account for in precision agriculture relative to agricultural inputs? For example, is the variation in nitrogen (N) mineralization within a field large enough relative to fertilizer N input to significantly reduce leaching losses of nitrate (NO_3^-) when we account for this variation in fertilizer application?
(3) What are the scales at which ecological factors influence the environmental performance of agricultural production systems? How do precision agriculture practices interact with these scales?

Ecological factors influencing the environmental performance of agricultural production systems

The environmental performance of agricultural production systems is fundamentally constrained by temporal discontinuities in nutrient cycling processes, high levels of soil disturbance, high levels of nutrient enrichment and a lack of resistance to the disturbance of extreme climatic events (e.g. high rainfall). In the sections below, I will elaborate on each of these constraints and analyse how precision agriculture techniques can or cannot affect these factors.

Temporal discontinuities

A major source of nutrient loss from agricultural production systems arises from the loss of plant 'control' over nutrient cycling caused by harvest. Many studies have shown that leaching losses occur primarily in the fall and spring when there is not an active 'sink' for nutrients in the plant community (Gold et al 1990, Smith et al 1994, Addiscott et al 1991, National Research Council 1993, Fig. 1). This phenomenon is well studied in natural ecosystems as well. Harvest (e.g. 'clear cutting') of forest ecosystems has been shown to lead to high rates of nutrient loss until biotic (i.e. plant) control over hydrological and nutrient cycling processes is re-established by regrowth (Bormann & Likens 1979). Arable agricultural production systems are essentially 'clear cut' at least once a year, leaving them inherently vulnerable to high nutrient loss. This vulnerability is exacerbated by the fact that in many areas, plants are removed just at the beginning of the season with the highest potential for hydrological losses of nutrients.

Precision agriculture techniques that vary the amount of fertilizer input based on variation in yield potential or nutrient availability do not have a high potential to

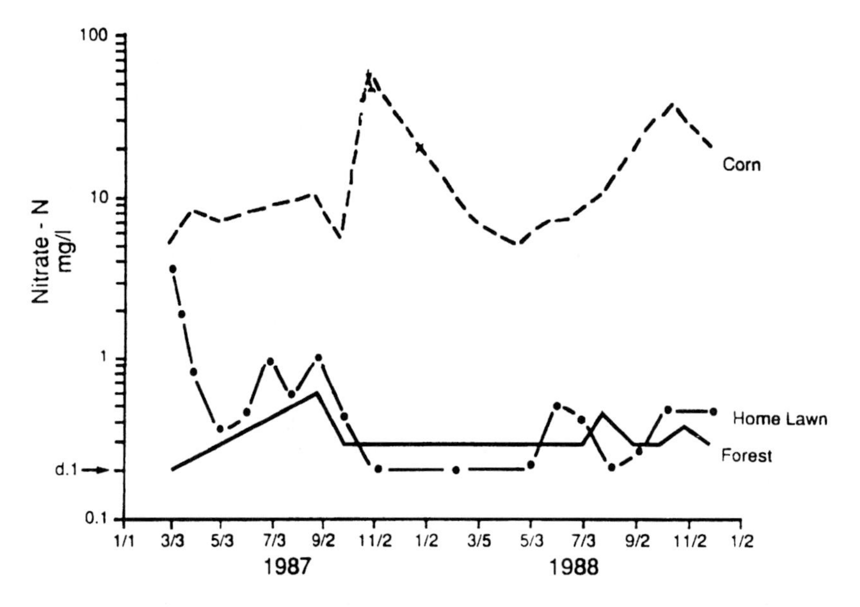

FIG. 1. Nitrate concentrations in groundwater leachate from a maize (corn) agricultural production system, a home lawn (grass) and oak-dominated forest on well drained silt loam soils in Rhode Island, USA. Adapted from Gold et al (1990).

reduce nutrient losses arising from temporal discontinuities. These losses are caused by a structural deficiency in the agricultural system and thus should not be very sensitive to spatially variable fertilizer input. Many studies have shown that the nutrients that are lost in the autumn, winter and spring seasons are not 'left over' fertilizer from the previous spring or summer application (Addiscott et al 1991, Magdoff 1991, Shipley et al 1992, Davies & Sylvester-Bradley 1995, Table 1). Rather, they are nutrients that have passed through plants and/or soil organic matter during the growing season and are lost due to a lack of a 'sink' in the plant community during autumn, winter or spring. Losses due to temporal discontinuities can only be ameliorated by altering the structural deficiencies of agricultural production systems through the use of

TABLE 1 Total residual and fertilizer-derived inorganic N in a maize agricultural production system fertilized with 168 kg N ha^{-1} in Maryland, USA

	Inorganic N (kg N ha^{-1})	*Percentage of total residual organic N*
Total residual N	87	100
Fertilizer-derived residual N	17	20

Fertilizer was applied in June, soil was sampled to a depth of 80 cm immediately after corn grain harvest in autumn. Data from Shipley et al (1992).

winter 'catch crops' (Addiscott et al 1991, Hargrove 1991), multiple or relay cropping systems (Vandermeer 1988, Kirschenmann 1991), or by landscape scale nutrient sinks (e.g. riparian buffer zones, National Research Council 1993).

It is interesting to note that many natural systems have a potential for high nutrient losses via temporal discontinuities. For example, in deciduous forests, leaves are lost during autumn, when the soil is still warm enough to support nutrient mineralization and loss processes, and are not re-grown until well after the soil warms in the spring. These systems do not suffer excessive nutrient losses, however, due to the presence of spring ephemeral vegetation (Muller & Bormann 1976) and perhaps more importantly, to the fact that forest litter has a high C : N ratio. As a result of this high C : N ratio, when leaves are added to the forest floor, net mineralization and loss are low (Fig. 1). In agricultural systems, where residues are N rich, mineralization is rapid, and potential and actual losses are high. It is also interesting to note that enrichment of forest ecosystems with N leads to changes in litter C : N ratio and increases in N loss (see below).

Soil disturbance

Disturbance of soil by tillage and planting activities is a fundamental constraint on the environmental performance of agricultural production systems because it exacerbates temporal discontinuities in nutrient cycling processes, creates small-scale spatial discontinuities in these processes, and increases the susceptibility of agricultural systems to soil erosion. None of these phenomena are ameliorated by precision agriculture techniques.

Tillage exacerbates temporal discontinuities by stimulating microbial decomposition and mineralization of residues at a time when plants are not present as a nutrient sink. Tillage can be an especially important source of nutrient loss if done in the autumn and not followed by sowing of a winter crop (Renard & Foster 1983, Addiscott et al 1991). Tillage creates spatial discontinuities in nutrient cycling processes by mixing the soil and disrupting links between plant roots, mycorrhizae and soil microbes. These links facilitate rapid uptake of mineralized nutrients by plants or microbes, preventing hydrological loss (Newman & Eason 1989). Tillage increases the susceptibility of soil to erosion by disrupting soil structure and removing residues (Renard & Foster 1983). Soil disturbance, like harvest, is a structural deficiency in agricultural production systems that is not addressed by spatially variable application of inputs. Rather, this problem can only be addressed by structural manipulation of the system through reduced tillage, increased residue cover or multiple cropping systems (National Research Council 1993).

Nutrient enrichment

It is clear that nutrient enrichment is a necessary component of agricultural production. However, studies of nutrient cycling and loss processes in natural ecosystems show that the need to add nutrients to agricultural production systems is a fundamental constraint on their environmental performance, and that this constraint is not very amenable to amelioration by precision agriculture techniques.

Studies of natural ecosystems, especially temperate forests, have shown that nutrient losses from these systems are surprisingly sensitive to variation in nutrient input (Aber et al 1989, Johnson 1992, Fig. 2). For N, most forest ecosystems appear to have a threshold amount of N that they can absorb (e.g. 5–20 kg N ha^{-1} y^{-1}). Above this threshold, losses increase linearly with input (Aber et al 1989, Johnson 1992). These studies are sobering because the level of input that these forests are responding to is much lower than the level of input common in agricultural systems. Moreover, forest ecosystems should have a higher threshold ability to absorb nutrients than agricultural systems due to their inherently lower nutrient status, higher soil organic matter levels and permanent plant cover. Although spatially variable fertilizer application practices attempt to account for the inherent ability of a cropping system to use N, these studies of forest N 'saturation' suggest that unless fertilizer inputs are reduced dramatically (to a point where crop yield would likely be unacceptably low), losses will still be high. These studies also suggest that precision variation of nutrient inputs will not ameliorate the fundamental constraint that nutrient enrichment places on the

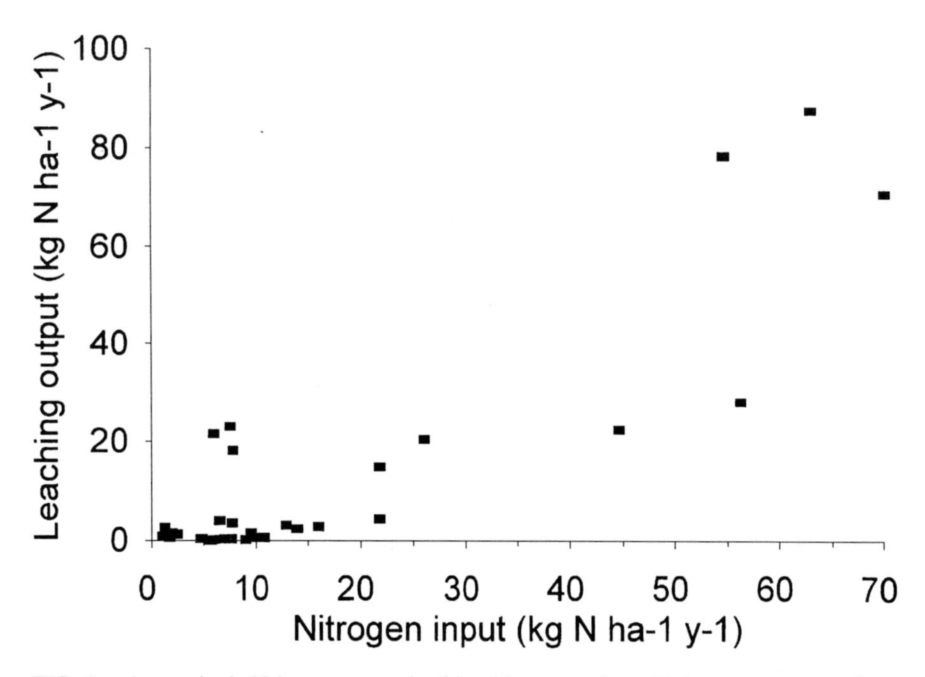

FIG. 2. Atmospheric N inputs versus leaching N outputs from 31 forest ecosystems. Data from Johnson (1992).

TABLE 2 Soil loss associated with total and low-return-frequency (large event) rainfall events from 1972–1991 in a soybean/grain/sorghum agricultural production system in Georgia, USA

	No. of storms	*Rainfall (mm)*	*Soil loss (Mg ha^{-1})*
All storms	934	22205	64.6
[a]LRF storms	11	1195	27.0
LRF storms (%)	1.2	5.4	41.8

[a]LRF, low return frequency storms.
Data from Langdale et al (1992).

environmental performance of agricultural production systems. Rather, structural changes that improve plant nutrient-use efficiency, greater harvest output of nutrients, or landscape-scale nutrient sinks (e.g. riparian buffer zones) will be required to address this problem.

Resistance to extreme climatic events

Studies of both natural and managed ecosystems have shown that extreme climatic events, particularly large rainfall events, are critical threats to the long-term sustainability of ecosystems (Bormann & Likens 1979, Edwards & Owens 1991). These events are frequently responsible for a major portion of the nutrient and sediment loss from many ecosystems (e.g. less than 5% of the storms can produce more than 50% of the sediment loss that occurs from a system over a multi-year period; Langdale et al 1992, Table 2). Agricultural systems are poorly designed to 'resist' the stress of these events. This lack of resistance is another fundamental constraint to the long-term environmental performance and sustainability of agricultural production systems that is not ameliorated by precision agriculture techniques. It is important to note that many natural ecosystems, especially forests, have multiple structural components that increase ecosystem resistance to extreme climatic events. In temperate forests, rainfall energy is dispersed by a multi-layered plant canopy (even in winter) and by a forest floor rich in organic matter. High levels of soil organic matter also allow for well-structured soils that facilitate infiltration in these systems. Most forest ecosystems have extremely low rates of soil erosion (less than or equal to the natural rate of soil formation) even in areas with abundant, high energy rainfall (Bormann & Likens 1979). Incorporating multiple layers of resistance to erosion is necessary to reduce the risk that extreme rainfall events pose to the long-term environmental performance and sustainability of agricultural production systems (National Research Council 1993).

Is the variation in ecological processes that we exploit in precision agriculture big enough relative to inputs to affect the environmental performance of agricultural production systems?

One of the main ideas in precision agriculture is that of accounting for natural variation in ecological variables in the application of agricultural inputs. There is particular interest in adjusting fertilizer N applications to variation in the natural N supplying power (e.g. mineralization capacity) of the soil within a field. However, for these adjustments to be effective in reducing N loss from agriculture, the magnitude of variation in mineralization must be significant relative to the fertilizer input for there to be a significant reduction in residual, i.e. leachable N.

Rates of N mineralization have been measured in many studies of natural and agricultural ecosystems, using many different techniques. In temperate forests and grasslands, N mineralization can be expected to provide 50–100 kg N ha^{-1} y^{-1} to plants (Schlesinger 1991). In agricultural soils, background mineralization is expected to be considerably less than in natural ecosystems due to the lower levels of soil organic matter, particularly 'active' soil organic matter in cultivated soils (Paul & Clark 1996). These data suggest that the magnitude of the variation in N mineralization that we hope to manage with precision agriculture is small relative to fertilizer input (150–250 kg N ha^{-1} y^{-1}), and that accounting for this variation will not reduce residual soil NO_3^- levels and loss to a significant extent.

Several studies that have analysed spatial variation in biological processes and soil NO_3^- in agricultural fields support the idea that the variation in biological processes is small relative to fertilizer inputs. In a fertilized field in Iowa, USA, Cambardella et al (1994) found that pools of 'mineralizable' and microbial biomass N averaged 2 and 10 g N m^{-2}, which are small relative to fertilizer inputs which

TABLE 3 Descriptive and spatial statistics for microbial biomass N (MBN), mineralizable N (MIN), denitrification (DEN) and soil NO_3^- from a maize/soybean agricultural production system in Iowa, USA

	Mean	*Minimum*	*Maximum*	[a]*Nugget (%)*	[b]*Range (m)*	[c]*Class*
MBN (g m^{-2})	13.3	0.41	39	52	30	M
MIN (g m^{-2})	1.9	−34.8	54.2	57	38	W
DEN (g m^{-2} d^{-1})	0.105	0.003	0.764	56	75	M
NO_3^-–N (g m^{-2})	4.95	0.20	20.2	79	201	W

The system was sampled just before maize planting in April 1992, at 241 locations in a 250 m × 250 m grid with points separated by 2, 5, 10 or 25 m intervals. Data from Cambardella et al (1994).

[a]Nugget = (nugget semivariance/total semivariance) × 100. This is an index of the percentage of the spatial variance that occurs at less than 2 m.

[b]Range = the range of spatial dependence. Samples closer than this range are spatially related.

[c]Class is an index of spatial dependence: S, strong (nugget <25%); M, moderate (nugget between 25 and 75%); W, weak (nugget >75%).

range from 15–25 g N m^{-2} in their region (Table 3). Denitrification, a process that consumes soil NO_3^-, averaged only 0.1 g N m^{-2} d^{-1}. Interestingly, the range of variation in the point measurements of these processes was much more impressive. Mineralizable N ranged from −34 to 54 g N m^{-2} and denitrification ranged from 0.003 to 0.764 g N m^{-2} d^{-1}. However, much of this variability occurred at scales too small to be accounted for by spatially variable fertilizer application (see below). Most importantly, there was little correlation between spatial patterns of biological variables and spatial patterns of soil NO_3^-, suggesting that these variables do not influence levels of residual NO_3^- (Fig. 3).

Similarly to Cambardella et al (1994), Robertson et al (1993, 1997) and Gross et al (1995) also found little correlation between spatial patterns of soil biological processes and NO_3^- levels (Figs 4, 5). These investigators compared natural (old fields and forests) and agricultural (maize, soybeans) ecosystems in Michigan, USA, and found that the agricultural fields had relatively high and uniform soil NO_3^- levels relative to the natural ecosystems, reflecting control by extrinsic factors such as fertilization and cultivation rather than by internal biological processes. Ferguson & Hergert (1996) found that spatially variable fertilizer application had no effect on mean residual soil NO_3^- concentrations in an irrigated maize field in Nebraska, USA.

Do the scales of variation in ecological processes match the scales of precision agriculture practices?

There are two scale congruence questions that need to be addressed in an analysis of the effects of ecological processes on the ability of precision agriculture to improve the environmental performance of agricultural production systems. First, we need to determine whether ecological processes vary on a scale amenable to management. For example, if there is significant variation in these processes at very small scales (cm), it will not be possible to exploit this variation with precision management. Second, we need to determine whether precision agriculture techniques exploit what is already known about spatial variation in soil and ecological processes. For example, extensive work has gone into analysis of functional differences between different soil types (polypedons) and into generation of pedotransfer functions to characterize the environmental performance of different soil types (Bouma 1989). It is not clear that the systematic sampling used in precision agriculture can exploit this knowledge base.

Several studies have shown that a variety of ecological processes vary at scales much smaller than those amenable to precision agriculture techniques. In a fertilized field in Iowa, USA, Cambardella et al (1994) found that mineralizable N and microbial biomass N showed spatial dependence up to a range of 30 and 38 m, while soil NO_3^- levels showed weak spatial dependence, at a range of 200 m (Table 3). Robertson et al (1997) found that most soil biological variables showed spatial dependence at a range of approximately 40 m, while soil NO_3^- levels and crop production showed spatial

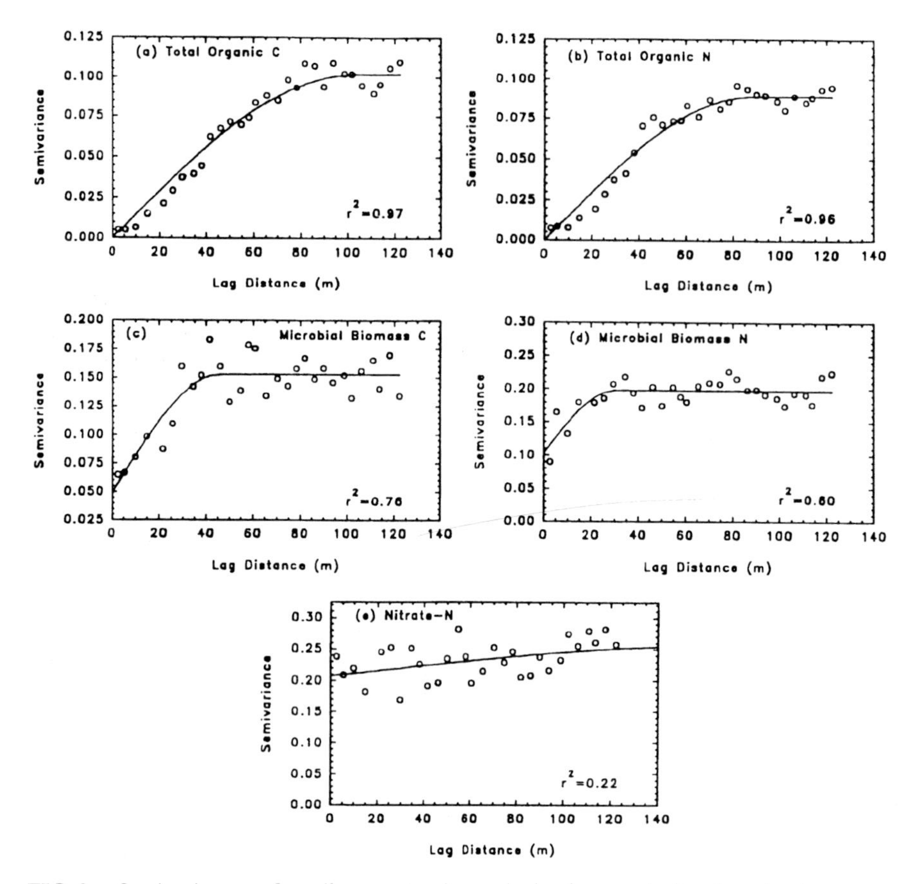

FIG. 3. Semivariograms for soil parameters in a maize/soybean agricultural production system in Iowa, USA. The system was sampled just before maize planting in April 1992, at 241 locations in a 250 m × 250 m grid with points separated by 2, 5, 10 or 25 m intervals. From Cambardella et al (1994).

dependence at much larger scales (approximately 90 m). The biological processes had high 'nugget' variance, with a significant portion of their variation occurring at scales less than 5 m. Studies in other, natural (Smith et al 1994, Schlesinger et al 1996) and cultivated (Rochette et al 1991) ecosystems have also found very small (<5 m) ranges of spatial dependence for soil biological properties. As discussed above, this small-scale variation in biological processes can be extreme (e.g. N mineralization or denitrification can vary from 0–1 $g\,N\,m^{-2}\,d^{-1}$ at the cm scale; Parkin 1987). Unfortunately, at scales relevant to even precision management (5–10 m), this variation is much less dramatic and coherent.

In addition to the high inherent variance of ecological processes at small scales there are plant- and fertilizer-driven patterns of spatial variation that are difficult to account

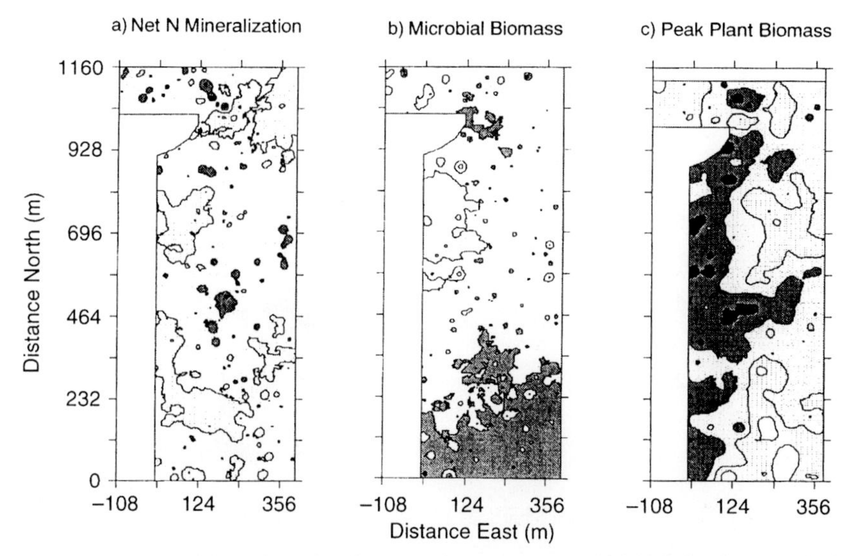

FIG. 4. Isopleths for selected soil properties across a maize/alfalfa/soybean agricultural production system in Michigan, USA. The system was sampled just before soybean planting in May 1992 at 192 locations in a stratified unaligned grid design. Patterns denote even increments in levels for a given property across its range: (a) net N mineralization (0.30–0.70 mg N kg^{-1} d^{-1}); (b) microbial biomass (39–95 mg C kg^{-1}); (c) peak plant (soybean) biomass (170–347 g C m^{-2}). From Robertson et al (1997).

for with precision agriculture techniques. A relatively large body of work has gone into describing the role of plants in structuring spatial dependence of soil biological processes in natural ecosystems (Robertson et al 1993, Jackson & Caldwell 1993, Caldwell & Pearcy 1994, Gross et al 1995). Other studies have found marked differences in soil biological processes between within-row and inter-row sampling locations (Klemedtsson et al 1991). Application of fertilizer in bands, which creates extremely high concentrations of soil NO_3^- in very narrow areas, can be a dominant factor influencing spatial patterns of NO_3^- leaching (Gorres 1992).

Given the spatial incongruence between soil biological processes and precision agriculture techniques, it is important to consider whether it is possible to reconcile these techniques with the extensive research base on functional characterization and classification of mapped soil types (i.e. polypedons and pedotransfer functions). It is not clear that it is logical to establish a systematic sampling grid, as is done in precision management, in a field with readily discernible patterns of topography and soil type. Many studies have established systematic relationships between soil type and a wide range of soil biological, chemical and physical processes (Bouma 1989). It would be useful to determine whether these processes vary with crop yield in a systematic way. To the extent that soil type integrates a range of factors that influence plants, combining knowledge of soil functional differences with precision yield mapping techniques may be a useful refinement to precision management techniques.

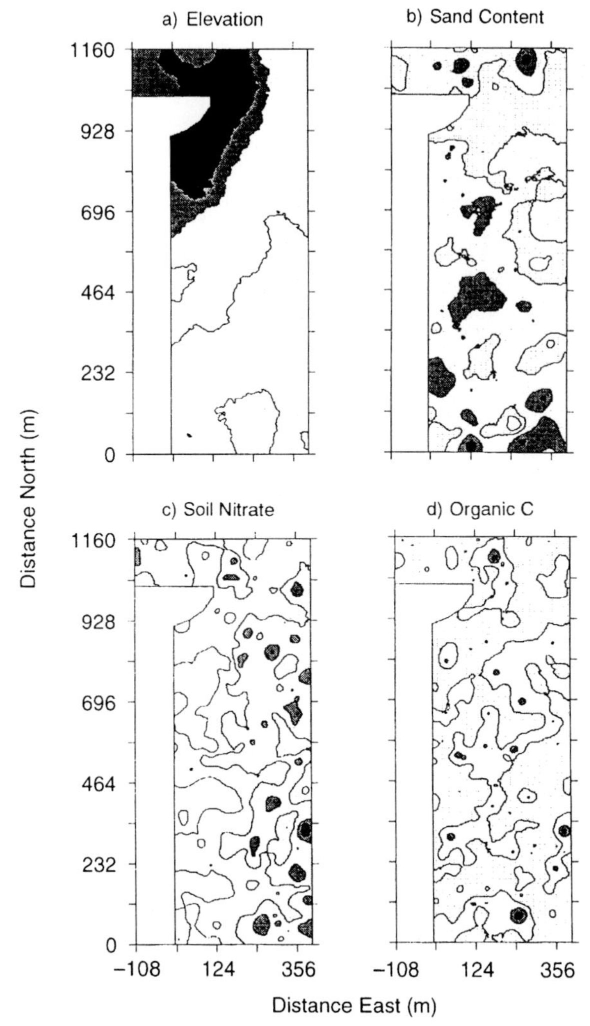

FIG. 5. Isopleths for selected soil properties across a maize/alfalfa/soybean agricultural production system in Michigan, USA. The system was sampled just before soybean planting in May 1992 at 192 locations in a stratified unaligned grid design. Patterns denote even increments in levels for a given property across its range: (a) topographic relief (285–291 m); (b) sand content (35–63% sand); (c) soil nitrate (5.7–15.1 mg N kg^{-1}); and (d) soil organic C (0.85–1.93% C). From Robertson et al (1997).

References

Aber JD, Nadelhoffer KJ, Steudler P, Melillo JM 1989 Nitrogen saturation in northern forest ecosystems. BioSci 39:378–386

Addiscott TM, Whitmore AP, Powlson DS 1991 Farming, fertilizers and the nitrate problem. CAB International, Wallingford

Bormann FH, Likens GE 1979 Pattern and process in a forest ecosystem. Springer-Verlag, New York
Bouma J 1989 Using soil survey data for quantitative land evaluation. Adv Soil Sci 9:177–213
Caldwell MM, Pearcy RW (eds) 1994 Exploitation of environmental heterogeneity by plants. Academic Press, New York
Cambardella CA, Moorman TB, Novak JM et al 1994 Field-scale variability in soil properties in central Iowa soils. Soil Sci Soc Am J 58:1501–1511
Davies DB, Sylvester-Bradley R 1995 The contribution of fertiliser nitrogen to leachable nitrogen in the UK: a review. J Sci Food Agric 68:399–406
Edwards WM, Owens LB 1991 Large storm effects on total erosion. J Soil Water Cons 46:75–80
Ferguson RB, Hergert GW 1996 Variable application of nitrogen and nitrapyrin on coarse-textured soils. Agron Abstr 88:314
Gold AJ, Deragon WR, Sullivan WM, Lemunyon JL 1990 Nitrate–nitrogen losses to groundwater from rural and suburban land uses. J Soil Water Cons 45:305–310
Gorres JH 1992 Characterization and simulation of spatial patterns of nitrate–nitrogen leaching from row-cropped silage corn as affected by fertilizer uniformity. MS Thesis, University of Rhode Island, Kingston, RI, USA
Gross KL, Pregitzer KS, Burton AJ 1995 Spatial variation in nitrogen availability in three successional plant communities. J Ecol 83:357–367
Hargrove WL (ed) 1991 Cover crops for clean water. Soil and Water Conservation Society, Ankeny, IA
Jackson RB, Caldwell MM 1993 The scale of nutrient heterogeneity around individual plants and its quantification with geostatistics. Ecology 74:612–614
Johnson DW 1992 Nitrogen retention in forest soils. J Environ Qual 21:1–12
Kirschenmann F 1991 Fundamental fallacies of building agricultural sustainability. J Soil Water Cons 46:165–168
Klemedtsson L, Simkins S, Svensson BH, Johnsson H, Rosswall T 1991 Soil denitrification in three cropping systems characterized by differences in nitrogen and carbon supply. Plant Soil 138:273–286
Langdale GW, Mills WC, Thomas AW 1992 Use of conservation tillage to retard erosive effects of large storms. J Soil Water Cons 47:257–260
Magdoff F 1991 Understanding the Magdoff pre-sidedness nitrate test for corn. J Prod Agric 4:297–305
Muller RN, Bormann FH 1976 Role of *Erythronium americanum* Ker in energy flow and nutrient dynamics in the northern hardwood forest. Science 193:1126–1128
National Research Council 1993 Soil and water quality: an agenda for agriculture. National Academy Press, Washington DC
Newman EI, Eason WR 1989 Cycling of nutrients from dying roots to living plants, including the role of mycorrhizas. Plant Soil 115:211–216
Parkin TB 1987 Soil microsites as a source of denitrification variability. Soil Sci Soc Am J 51:1194–1199
Paul EA, Clark FE 1996 Soil microbiology and biochemistry, 2nd edn. Academic Press, New York
Renard KG, Foster GR 1983 Soil conservation: principles of erosion by water In: Soil conservation, dryland agriculture. American Society of Agronomy, Madison (Agronomy Monograph 23)
Robertson GP, Crum JR, Ellis BG 1993 The spatial variability of soil resources following long-term disturbance. Oecologia 96:451–456
Robertson GP, Klingensmith KM, Klug MJ, Paul EA 1997 Soil resources, microbial activity, and primary production across an agricultural ecosystem. Ecol Appl 7:158–170

Rochette P, Desjardins RL, Pattey E 1991 Spatial and temporal variability of soil respiration in agricultural fields. Can J Soil Sci 71:189–196

Schlesinger WH 1991 Biogeochemistry: an analysis of global change. Academic Press, New York

Schlesinger WH, Raikes JA, Hartley AE, Cross AF 1996 On the spatial pattern of soil nutrients in desert ecosystems. Ecology 77:364–374

Shipley PR, Meisinger JJ, Decker AM 1992 Conserving residual corn fertilizer nitrogen with winter cover crops. Agron J 84:869–876

Smith JL, Halvorson JL, Bolton JF 1994 Spatial relationships of soil microbial biomass and C and N mineralization in a semi-arid shrub-steppe ecosystems. Soil Biol Biochem 29:1151–1159

Vandermeer JH 1988 The ecology of intercropping. Cambridge University Press, New York

DISCUSSION

Stafford: Concerning your temporal discontinuity argument, you showed that residual nutrients are spatially variable. Surely, in the development of a fertilizer application strategy, one of the factors to include is the residual nutrient variability. If the application strategy is based partly on residual nutrient over a series of years, then you will see a benefit over and against uniform application of fertilizer.

Groffman: One would hope so. It all depends on what leads to the production of residual nutrient content. Over a series of years you could reduce the residual nitrate concentrations through precision application. The concern I have is that the residual nitrate is not 'left over' fertilizer. Controlling residual nitrate depends on our ability to influence nitrogen cycles through organic matter, plants and microbial biomass, which ultimately control the levels of residual N. We should be able to influence these, but it is a complex cycle that hasn't proved particularly amenable to management.

Rabbinge: That's a good point. One has to take into account the whole mineral balance, be it from organic manure and matter or artificial fertilizer.

Kropff: I agree that it is important to stress the variability in residual nutrients and that the processes involved are complex. In your study you tried to correlate the residual nitrogen with environmental factors such as organic matter in a static way, whereas Johan Bouma tried to model nitrate losses using mechanistic models (Bouma 1997, this volume). Would you expect that with such mechanistic models you would find similar patterns in your situation?

Groffman: Johan was modelling variation of production, and the patterns of variability within the field were relatively stable. All farmers know this: they are aware that they have both good and bad spots in their fields. What I'm arguing is that you can account for those good or bad spots in a field, but fundamentally you are going to have a problem at the end of the season because there will be a lot of nitrogen left that's going to leach out. Nothing I said contradicts anything that Johan Bouma said.

Kropff: Nitrogen losses and yield were predicted quite accurately by the model. With such information it will be possible to manage the field in such a way that bad spots are taken account of and the nitrogen losses are minimized.

Groffman: It comes down to the question of fine-tuning versus structural changes. His system is very finely tuned, but the question remains: is that enough? To the extent that the fine-tuning doesn't address some big problems, it may not be.

Bouma: I would like to recount the story of a visit we made about 10 days ago to one of our farmers. We have been working with this farmer in southern Holland and we visit him regularly. We asked him what was the most crucial thing he would like to know about the way his land performed, and he said that it was the dynamics of the nitrogen in relation to organic matter content and the type of organic matter. I was intrigued by that, because when you mentioned that 20% is mineral nitrogen and 80% is something else, this is not essential—nitrogen is nitrogen. In models on nitrogen transformations there are rate constants for various pools, and we're finding that in the very same soil series in the Netherlands we have a soil that has been farmed by biological dynamic methods for 70 years. We have another farm, highly mechanized, and we have an old meadow. The rate-decomposition constants of organic matter have been measured for each management type. They are put in the models and management decisions on nitrogen fertilization can be based on model runs. Without belittling the difficulties I am fairly convinced that we can do this. The soil structure is also important: sometimes you find that the organic matter is in natural soil aggregates and is thus inaccessible, and therefore doesn't decompose the way it is supposed to. This can be taken into account. One of the aspects I find fascinating about the modelling is that we look at the process in the context of all management activities on a farm. We factor in not only fertilization, but also tillage, planting, weeding and so on. I hope we can move away from the sorts of discussions we occasionally have with ecological farmers about one form of nitrogen being better than another. Different forms have different decomposition constants and they can be measured (Droogens & Bouma 1997).

Goense: As an agricultural technician, I was struck by one of your later remarks where you referred to the effect of banding of nitrogen. I interpret these rather erratic non-correlated nitrogen figures to be the microvariability which occurs on such a small scale that we do not deal with it on a crop basis with applied statistical techniques. I assume that what you mean by the effects of 'banding' is that this might cause some of the unpredictable nitrogen data.

Second, you mentioned the effect of individual plants causing variability. There are some attempts from an engineering point of view to guide machinery so precisely that you know exactly where individual plants are located, and application occurs in relation to that microvariability. Would this be a means of precision agriculture that would work?

Groffman: With regard to the banding and the plant row–inter-row differences, I'm suggesting that these are factors that influence nitrate leaching that are perhaps not very amenable to improvement with precision agriculture. In precision agriculture, we think about what controls the variability in nitrate leaching in a field — a spot with coarse texture, for instance. But we know that there's also a lot of smaller-scale variability, for example that caused by fertilization bands and plant roots. If we don't deal with these in precision agriculture, we're not going to make much progress in reducing nitrate leaching. One of the main objectives of my paper was to get us to think of the complete set of factors that influence the variability of leaching, so that we can perhaps account for them with precision management.

Rabbinge: To follow up on what you were saying, my impression is that you are arguing that there is a lot of nitrate buffering and smoothing capacity within the soil. If you consider how much nitrate there is available within the organic matter in relation to the amount of fertilizer you are adding, there is a difference in the order of magnitude of 10- or 200-fold. This means that a small change in mineralization can completely over-ride the small bit you are adding through your fertilizer.

Groffman: I'm saying the opposite of that — the mineralization is small compared with the fertilizer application.

Kropff: The mineralization rate on good agricultural soils during the growing season ranges between 0.5 and 2 kg N/ha, which is quite low compared with fertilizer applications of up to 200 kg N/ha for a whole season. However, on a whole-season basis they are both important, and mineralization is very important at the end of the growing season, because the process of mineralization continues. Therefore the whole complex of processes that determine nitrogen availability and losses has to be studied and models can be quite helpful for studying these complex processes.

Besides site-specific nitrogen application to avoid losses at the end of the season, losses can be minimized by using nitrogen in a biological way with a second crop. One of the main developments in ecological agriculture is the use of catch crops. In this system there is an undercrop which starts growing rapidly directly after harvest of the main crop, which then takes up the available nitrate and fixes it in biomass that can be ploughed under later on. This is a good way of carrying your nitrogen over to the next year, thus avoiding the losses caused by leaching at the end of the growing season.

McBratney: Peter Groffman, you are essentially talking about microbial ecology, and your proposition is that it has little effect. Does this actually make the practice of precision agriculture easier with regard to fertilization? If there was a stronger interaction it might make it more difficult.

Groffman: The question I am asking is: do you really need to come up with some clever way to measure the spatial variability of N mineralization in the field? My hypothesis is that you don't. I know there are students at universities in the USA who are measuring spatial variation in N mineralization across the field and seeing whether it matters. The preliminary data suggest that it doesn't. We have talked about spatial constancy in time: the two examples I showed are from a single sample date, and so we don't know how stable the residual nitrate concentrations are, or how stable those mineralization rates are. In answer to your question, yes, theoretically it could make precision agriculture much easier if you didn't have to bother with all the microbial ecology.

Webster: Whitmore et al (1983) determined the spatial variation in the rate of nitrogen mineralization at Rothamsted and found that the correlation range is very short — between 2 and 4 m. If you wish to distinguish and map variation at that scale, then the sampling task is massive and taking it into account in precision agriculture would be enormously difficult. If on the other hand you decide that you cannot expect to resolve the pattern at that scale and there is no coarser component to the pattern, then the practical solution is straightforward — uniform application of fertilizer.

Groffman: For example, I showed spatial variation in mineralizable nitrogen going from -30 to $+50\,g/m^2$, but this variation was at the scale of centimetres. So there is great variability, but at such a small scale that it is not amenable to precision management. The question of at what scale there is coherent variation in mineralization is an important one. It may be centimetres, or 2–4 m, or 10–20 m. It may vary by soil or crop type. My hypothesis is that the really big variation occurs at scales too small to address with precision management.

References

Bouma J 1997 Precision agriculture: introduction to the spatial and temporal variability of environmental quality. In: Precision agriculture: spatial and temporal variability of environmental quality. Wiley, Chichester (Ciba Found Symp 210) p 5–17

Droogens P, Bouma J 1997 Soil survey input in exploratory modelling of sustainable land management practices. Soil Sci Soc Amer J, in press

Whitmore AP, Addiscott TM, Webster R, Thomas VH 1983 Spatial variability of soil nitrogen and related factors. J Sci Food Agriculture 34:268–269

General discussion I

Rabbinge: I have selected four subjects for this general discussion. The first is related to Pascal Monestiez's paper and concerns the application of the techniques and models that have been developed to account for sparse or incomplete data. How can we apply them in a way which addresses the problems we encounter in precision agriculture?

The second question relates to the use of iterative procedures in the development of these techniques. This iterative process may in itself be useful, but how is that related to the final decisions which the farmer has to take?

The third discussion topic is related to Peter Groffman's paper. He raised the question of the fundamental constraints on the ecological performance of agricultural production systems. This is a crucial question which I think we should discuss in detail, because he demonstrated clearly that there are temporal discontinuities but that these are minor problems in relation to the fundamental problems which we address. If that is the situation then precision farming is taking care of the margins but it is not addressing the main effects.

The final problem concerns the magnitude of the ecological processes. What is the scale of the heterogeneity? Does it operate at a scale of centimetres, metres or even hundreds of metres?

Burrough: In these ecological conditions for crop growth which we have discussed, I've not heard anyone talk about solar radiation inputs. In hilly countries, and certainly those that are a fair distance north and south of the equator, this could be important, especially for perennial crops. Given good digital elevation models, this is something we can calculate with geographical information systems (GIS) fairly easily. The other thing we could do with these elevation models is to calculate overland and subsurface flow of water, which is another input to give you local variation. One could combine temperature differences in site and also moisture supply purely from the geometry of the landscape.

Monestiez: The problem of solar radiation may be addressed simply with GIS and elevation models, but it's true that in most northern regions the solar radiation issue is more one of cloud cover than one of slope orientation and topography. This has to be addressed through teledetection and interpolated solar radiation data rather than GIS. If you integrate cloud cover over sufficiently long a time—the period of crop growth—it is more or less a random process.

Rabbinge: This is a problem of course. If you are considering crop growth you have to do it not only on a daily basis, but also the time steps and accuracy should be dictated by the many processes that occur. You may have responses to environmental conditions which differ widely in time coefficient. In this case it's not possible to integrate over a long period, because the processes we are studying have a much smaller time coefficient which is really very small in comparison with the processes occurring over the whole growing season.

Monestiez: From external information such as remote sensing images, you may calibrate or fit some geostatistical model, and use the fitted covariances for kriging with very few weather stations. This may be used, for example, to interpolate cloud cover with few solar radiation monitoring sites but long time series and several satellite images.

Rabbinge: Essentially, what you are saying is that if there is a scarcity of data we can't solve this problem with a statistical technique. Do you think that is really true? We have many sophisticated techniques that may overcome such problems.

Burrough: I don't agree with everything you say. Depending on how accentuated your relief is, the geometry will be more or less important. Obviously in flat landscapes it's fairly unimportant; instead, the absorption of the atmosphere takes a dominant role. But if you've got South facing slopes as distinct from North facing slopes, particularly for permanent crops such as grapes, then the integrated amount of direct radiation over the year varies enormously over a landscape (I've computed this and I can demonstrate it). On top of that you have then got the reflected radiation from the different hillsides, plus the horizon effect (sometimes you're in shadow sometimes you're not), plus the transmissivity of the atmosphere. I agree with you that this is where we could use the remote sensing information to back up the information we have on the ground. The methods for computing direct solar irradiance are quite easy to do, so it gives us a good basis to start from. We can then talk to the remote sensing people about getting the other information and we can see how far we can get with that.

Rabbinge: The main problem, taking into account of course the location and the geometry of the crop, is that these sorts of data are heavily dependent on the input information you get from various stations.

Bregt: Even in the European context it is quite difficult to get these data. We have been constructing a crop growth monitoring system for the European Community, for forecasting the expected yield at the end of the growing season every year. It turned out that for many European stations the radiation data are lacking. It took a lot of work to derive an easy method which was based on topography and the distance from the coast and so on, to create a usable data set.

Burrough: In southern Spain they have some very efficient methods of water harvesting, which are based on topology—they catch the run-off from the hills and store it subsurface. The geometry of the landscape is extremely important for providing the moisture for their local agriculture.

Mulla: There is currently a good deal of interest in terrain modelling, from both the hydrological and the management points of view. Many landscape processes, including crop growth, the activity of microorganisms and flow of water, relate quite strongly to the curvature of the terrain and the catchment size. There is tremendous potential in precision farming for incorporating more of the terrain modelling approaches, not just from the point of view of estimating meteorological inputs, but also with regard to the growth of the crop itself, the functioning of different growth stages, the activity of microorganisms and the leaching of nutrients. In my experience, whenever there is residual nitrogen at some point in a field, it is usually an indication that this zone has either reduced infiltration rates or a small catchment size.

Hydrology is a key factor, and unfortunately most practitioners of precision agriculture are not keyed into its importance. There has been tremendous interest in precision farming that focuses on the nutrient content and biological properties of soils and the yield of the crops, but we have yet to incorporate the hydrological aspects into our management systems.

McBratney: I wanted to mention a concrete example with digital elevation models applied to precision agriculture. In this case, stubble mulch was managed differentially in relation to the slope aspect to control soil temperature for enhanced germination in the spring. The angle of discs was varied in proportion to the slope aspect. On south-facing slopes the disc angles were set so the mulch was incorporated and on north-facing slopes stubble was retained on the surface by having discs set parallel to the stubble row. In intermediate slope aspects the angle was set in-between.

Burrough: Another application is looking for frost pockets, where cold air is moving down slope, which is a real-time application.

Rabbinge: I would like us to move on to discuss the iterative procedure Pascal Monestiez described in his paper.

Monestiez: It's difficult to know how far you can go in adding more information to the deformation model. I am afraid that if you go too far the only thing you will succeed in is to reject the assumption of the existence of the deformation. Often if you have too many data you end up rejecting the model. Somehow we have to include this model in the larger class of space transformations, for example, transformation from the 2D geographic space to a 3D, or more isotropic and stationary space. If the complexity of the spatial covariance of the variable increases, you have to increase the dimension of what you want to represent. It then becomes very difficult to control what you are modelling by the transformation.

Rabbinge: So, in effect this is an optimization problem. Adding more detail will incur the cost of more labour and result in only marginal improvement.

Monestiez: Simplicity is the key. If you get an over-complex model for explanation, you will lose all the benefits for prediction.

Rasch: I think the crucial point is that you considered the time series as a stationary one. The different measurements are considered as replications. I don't think this is often the case. Of course, if you measure each year at the same point it may be, but this would take a long time. This is one of the crucial assumptions, that if we discuss robustness we also must consider non-stationary or at least seasonal components during time.

Barnett: I feel rather strongly about this model. There is some transformation function f which is supposed to be almost a 'magic' function — it is intended to take out all time-local temporal variation; it is intended to take out all the time effect over the space variation. It seems a wonderful function and I cannot say that I have ever encountered a practical situation where a simple scalar of function has had this effect of removing all of the time variation. If it does, that is great and then I guess that models of this sort are going to be very valuable. If it does not work in the sort of problems that I have looked at, then I have to ask myself what might be needed to make the model just a little more complex in order to represent what is really going on. Therefore I take a slightly different view on this idea of extra data. Extra data give you the facility to understand extra structure — extra structure that is generally there and needs to be explained. Just to give you one simple example, you could, as an alternative in the sort of problems that were being described, have fitted times series models on a point-by-point (location-by-location) basis. You could have estimated the parameters in those, and then have done something about fitting those parameters over the spatial region. I have seen this done; it can work. You can even finish up by kriging the residuals, and then you have the best of all possible worlds! But I am extremely worried about what from my experience seems likely to be the relatively limited field of application of a model built on the premise that a simple scalar transformation will take out all time effect.

Monestiez: One answer might be to work with a limited time series. It is true that you can have a change in time and you will never have nice asymptotic behaviour in time for spatial structure. But I am confident that this model will work better over a short time sequence with dense spatial samples than in other instances where there are few spatial samples over a long time series.

Rabbinge: So this method works best within limited, well defined conditions. It is therefore a useful tool for elaborating information used in precision agriculture.

Rasch: I think that this procedure may also be handled in a sequential way: after each observation we decide whether to stop or continue. This is because it is typical sequential, and you can stop when you have sufficient information.

Rabbinge: I would like to move to the million dollar question posed by Peter Groffman, concerning the fundamental ecological constraints on the performance of precision agriculture systems. How much basic information is needed to make such systems worthwhile and give them a considerable added value?

Thompson: I have a perspective question on the whole topic. Because we're using the term precision agriculture, does that mean we're only looking at things that are site-specific on a small scale? It would seem useful to be able to apply precision agriculture at any scale, including looking at what is happening in the areas outside of a field, if that allows us to manage the field more effectively over the short or long term.

Groffman: In much of my talk I was defining precision agriculture in a narrow sense to encompass spatially variable application of inputs, in particular of fertilizer. If we take a larger definition of precision agriculture which includes variation in tillage systems and incorporation of other structural elements and plant components, then the conflict that I set up becomes less of a problem. There's no reason that precision agriculture and structural changes are incompatible; they could go on at the same time.

It also depends fundamentally on what the goals are. Where are the data on environmental improvements associated with the techniques that we've talked about? I would like to see a small catchment where people have instituted precision agriculture techniques and seen a reduction in the groundwater nitrate and pesticide concentrations. These sorts of data are extremely important, and improvements need to be evaluated relative to specific environmental needs. For example, getting groundwater N concentrations below 10 mg/l to achieve a drinking water standard is relatively difficult, but if we are talking about eutrophication problems in the Baltic or Chesapeake bay, then 4 mg/l N is a big problem, so it depends on what we need to achieve. We need to see what improvements we are getting, and we need to balance that with what problems we need to solve and what standards we need to meet.

Burrough: We do have some data for the Rhine. The areas in former East Germany had a much lower input of nitrogen than those from West Germany. If you look at the total diffuse nitrogen input in the surface waters there are big differences. So, by implication, if you reduce the nitrogen input by precision agriculture you would expect to get large reductions.

Groffman: Yes, but did they have lower agricultural production as well in East Germany?

Burrough: Yes.

Groffman: What we want is to retain production levels but lower the environmental impact.

Kropff: That is indeed the core problem. With respect to nitrogen, quite some progress has been made in The Netherlands in reducing inputs and losses while maintaining production levels. For example, in a project undertaken by several Dutch institutions to develop integrated farming systems, systems were developed with 38 pilot farms. These farmers reduced pesticide use by 90% and nitrogen inputs by up to 25%. They had tremendous reductions in inputs and only a slight decrease in production, as a result of the increased efficiency of resource use. In terms of net income these pilot farmers were better off, as well. These farmers used site-specific management only at the field level, as they sampled mineral nitrogen early in the season and based their fertilizer application on that. Further site specification could help to improve the efficiency.

Bouma: Peter Groffman related nitrate content to patterns obtained for biomass and organic mapping. Essentially those observations are static. We have done very little research on this, but if we take the soil moisture regime and see what the changes in moisture content and temperature have been, we get a much better explanation of the observed nitrate contents, compared with just looking at static values. So I'm making a plea for a dynamic appraisal of the system and then those correlations will be a lot better. I remember from my microbiology courses that all organisms occur everywhere and that they thrive when the conditions are good.

Groffman: That is true, and that is much of what I was saying. When we harvest a crop and we till the soil, the microbes are very happy, they have fresh residues and so there is vigorous microbial activity. This is why nutrient losses are so high in the transitions between the crops, and we need to deal with those transitions much more aggressively.

Rabbinge: I would like to move to the subject of the scale of ecological processes and the level of precision in precision agriculture. If the major effects occur with a very small time coefficient and on a very small scale, and the minor effects on a large scale, then it is almost impossible to address the problem of heterogeneity with precision techniques.

Mulla: I think we should keep in perspective what conventional agriculture is as opposed to precision agriculture. Conventional agriculture is the uniform management of very large areas; precision management is the subdivision of these fields into the subunit management zones. The key point is that we now have equipment in the field which is capable of varying seeds, tillage, fertilizer or herbicides at a scale of something like 20 m. The question for precision agriculture is: can we characterize the mean value of the property that we think is key for the application of fertilizer or seeds or water at that scale? What is the scale at which we can actually characterize the mean in these subunits? My feeling is that there may be various scales in variability, but by going to a system which inherently tries to manage for the mean for the subunits of fields, the efficiency of agriculture is going to improve. We should keep that in mind, even though we may still have some difficulty in characterizing the variability within those areas and accounting for it. We can do a much better job at that scale than we can when we're talking about uniform management.

Stafford: We can talk in principle about managing smaller units, but when it comes to identifying the scale at which we might be varying fertilizer application rate, there are so many restraints: there are the technological restraints on application, there are economic restraints and agronomic restraints — before we think about the spatial variability of the crop and soil processes. Can we identify a rationally based scale? In our studies on spatially variable herbicide application, where there are far fewer factors involved, we've identified 4–6 m as a rational scale on economic and agronomic grounds. With fertilizer, there are so many other inputs that it may not be possible to define a scale.

Rasch: It seems to me you have several purposes that need to be handled at the same time. With regard to the question of which scale to use, we can use a mixed scale, use a grid of a certain size and around the endpoints we can have a smaller scale and more dense observation in a nested grid. The area of the smaller scale should be optimized for the precision required. We are just trying to do this investigation in a forthcoming research project.

Stafford: But how do you identify that grid size?

Rasch: I think you have an area where you should sample and to reach a precision fixed in advance you need a number of points to fulfil the precision requirements. From the number of points and the area you can deduce the grid size.

Rabbinge: The difficulty here is we have here both a scientific problem (the grid size) and a practical problem (management by the farmer). For each of these problems an adequate solution is needed.

Groffman: David Mulla defines the scale of precision management at about 20 m. We have technology where we can vary fertilizer input within 20 m. So now let's look at a field of 20 m grids. If I take 10 samples within that 20 m grid and measure nitrogen mineralization, it's going to vary from maybe 10–500 kg N/ha per year with each little sample. But the mean of that 20 m plot may be 30, and the mean of the next 20 m plot may be 50. I don't think that I'm going to see much more than 20–30 kg/ha variation among the 20 m grids. In one way 20–30 kg/ha per year is a lot of variation, but if your fertilizer input is 200 kg/ha then I don't think you're going to get a real fundamental difference in residual nitrogen losses because the variation of mineralization is just too small relative to the amount of nitrogen input into the systems.

Kropff: It is very important to study the relation between scale and variability first. If variability occurs at a scale at which equipment cannot operate, new technology has to be developed. For example, when you talk about urine spots in grassland it only makes sense to apply precision agriculture if you have equipment that can monitor soil fertility (urea content) and that can fertilize /at such a small scale as well. At our institute (AB-DLO) a methodology has recently been developed for that purpose and the equipment is now produced by industry. Urine spots are detected and the liquid fertilizer applicator is de-activated at these places in the field.

Bouma: The point of reference is the traditional method of agriculture. To use a rather corny analogy, in traditional farm management we used to have 'all the animals in one barn' and now we separate 'the elephants from the cows from the mice'. We only focus on major differences among soils within a field, not the small ones. We have tried to generalize these relationships for each of 64 observations in our field to see what entities behave differently from year to year. We found only two units that showed significant differences among the years: the more sandy area and the more clayey area, but there is still quite a bit of variability within these sub areas which pedologists are getting excited about. There will be a new soil map with those two areas and the basis of that has been a 30 year simulation of crop growth.

Burrough: This is in fact is the general problem of how you remove non-stationarity from your data, and when do you perceive your non-stationarity as being generated by some signal that you can either model deterministically or can account for with other information. We've seen many very different ways of doing this. On the one hand there is the 'that's the top of the hill, that's the middle, that's the bottom kind of approach', and on the other hand there is the sophisticated, intellectual approach when you have other kinds of data. There's a whole range of techniques. What we need to know is when one technique is appropriate and when another technique is appropriate.

Rabbinge: We need a toolbox. In the toolbox various tools are available. Each should be used for their own purpose and in the way that they're effective.

Burrough: And for a farmer this needs to be put up in a way that is easy and intuitive to use.

Webster: I would like to come back to David Mulla's technological point. He has a machine that can operate on 20 m × 20 m plots. His question concerned how we get the information to feed this system. Bluntly, he has got to be able to sample every 10 m, otherwise he cannot get the information. The cost of getting soil information at this density is enormous.

The following example illustrates what we might achieve if we will tolerate a rather coarser resolution. It is from a commercial farm on which the farmer wanted to vary his application of fertilizer, in this instance phosphorus (P), according to the demand of the crop (winter wheat) and to the nutrient status of the soil. The field covers some 27 ha, and the farmer sampled it at 50 m intervals. He had the soil samples analysed

TABLE 1 (*Webster*) Summary statistics

	Variable	
Statistic	*Yield* ($t\,ha^{-1}$)	*P* ($mg\,l^{-1}$)
Mean	7.23	20.1
Minimum	4.3	5.0
Maximum	10.5	57.0
Standard deviation	1.24	12.38
Variance	1.5509	153.35
Skew	−0.21	0.92

for P, and he measured the yield of wheat grain using a recording combine harvester to give 113 data on each. The results are summarized in Table 1 (*Webster*).

You see substantial variation in both variables; the largest yield is more than twice that of the smallest, and there is a tenfold difference between the smallest and largest P concentration. The smallest values are near the deficiency threshold.

Let us now look at the relations between yield and nutrient status geostatistically. I computed the variograms using the computing formula:

$$\gamma(\mathbf{h}) = \frac{1}{2m(\mathbf{h})} \sum_{i=1}^{m(\mathrm{h})} z(\mathbf{x}_i) - z(\mathbf{x}_i + \mathbf{h})^2, \qquad (1)$$

where $z(\mathbf{x}_i)$ and $z(\mathbf{x}_i + \mathbf{h})$ represent the measured values of yield or P at the sampling points $\mathbf{x}_i$ and $\mathbf{x}_i + \mathbf{h}$, and $m(\mathbf{h})$ is the number of paired comparisons separated by the lag $\mathbf{h}$. I fitted models to them by weighted least squares approximation. Figure 1(*Webster*) shows the models that fitted best for yield and P, and Table 2 (*Webster*) lists them and their parameters.

I kriged the variables at 25 m intervals for blocks of 24 m × 24 m, assuming that the fertilizer spreader covers 24 m at each pass. Figure 2 (*Webster*) shows the resulting maps for yield and P.

The variograms of the two variables resemble one another apart from the curvature in that of P at the long lags, as do the patterns on the maps, and one is tempted to conclude that there must be a causal relation between the variables. Inspection shows, however, that the relation is negative—where the yield is large, the P content is small and vice versa, and the Pearson correlation coefficient r is −0.42. This might strike you as rather surprising. What seems to be happening is that the larger the yield is (and assuming this is consistent from year to year) the more P is removed in the crop and so the soil's store of P is depleted there. The farmer, however, had hitherto applied a uniform dressing of an amount derived from analyses of bulked soil from the whole field, or had not applied any recently because on average the soil seemed to contain enough.

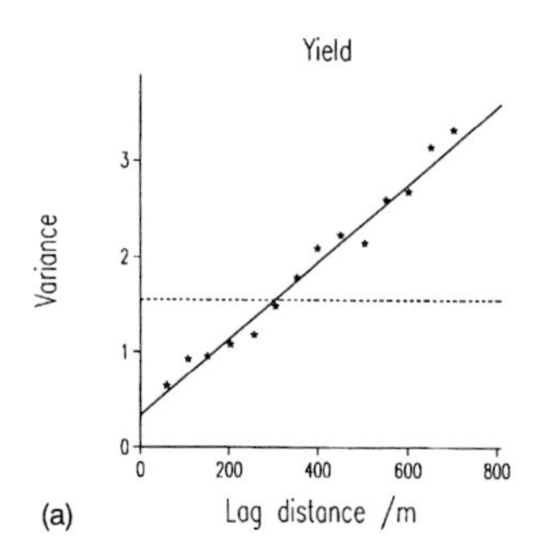

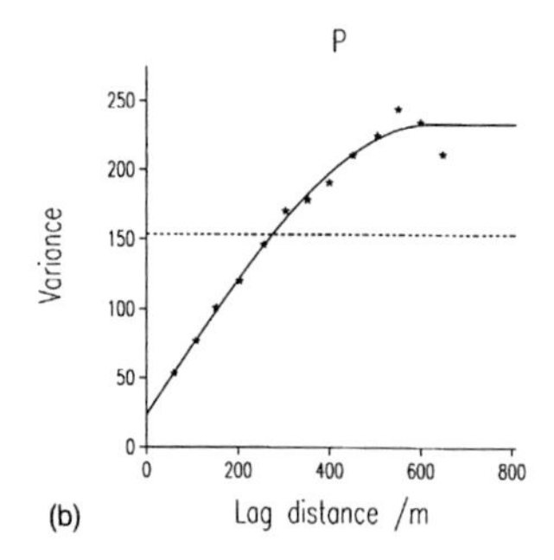

FIG. 1. (*Webster*) Variograms of (a) wheat yield and (b) phosphorus (P) concentration in the topsoil.

The available P is approaching the deficiency in parts of the field, yet is plentiful elsewhere. Here we have a clear case for differential application of fertilizer and the modern technology of precision farming. The spatial scale makes it feasible technologically. Only the economics conspire against it because of the need to analyse the soil for P from at least 100 sampling points, and the cost of that in any one year is more than the farmer can expect to gain in greater efficiency.

I thank Mr P. Chamberlain and Mr C. J. Dawson for providing the data for this illustration.

Groffman: Could you look at that field and pick that phosphate deficient spot?

Webster: No.

Groffman: I'm wondering whether we are making use of the useful work on polypedons and pedotransfer functions in precision agriculture. When I look at a field, there are five or six mapped soil types, and Johan Bouma's group in particular have established functional relationships between soil types and various functions related to water and nutrient dynamics. Are we using that information in precision management? We know a lot about functional variation within these fields just from the soil types. Why do we lay out a grid for precision agriculture that crosses over those boundaries? It seems that we're not utilizing the information we already have about functional variation within the field. Also, doesn't it violate some of the assumptions of the geostatistics if you cross over a boundary?

TABLE 2 (Webster) Models and coefficients of variograms

		Parameter			
Property	*Model*	c_0	c	a/m	m
Yield	Linear	0.3295			0.2016
P	Spherical	23.26	210.10	12.47	

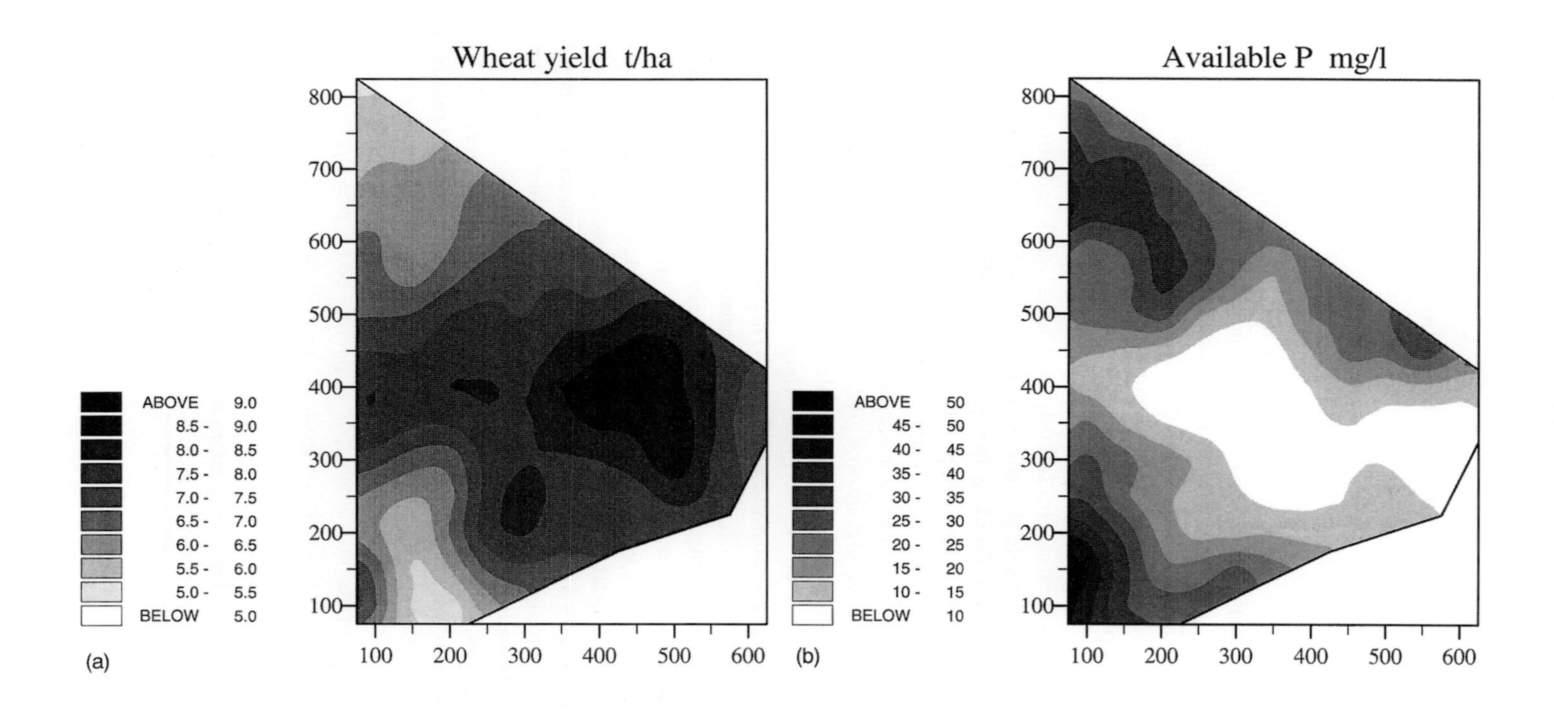

FIG. 2. (*Webster*) Maps of (a) wheat yield and (b) phosphorus (P) concentration in the topsoil.

Burrough: But the boundaries are not necessarily real; they are quite often arbitrary. They are things that people choose or decide on in committee rooms. Some of them are mapped in the field.

Bouma: The basic underlying philosophy of having genetic soil types is that it's morphometric: you can look at soils and you can put names on them. When we talk about functionality it's a different story. There are identical soil types from the genetic point of view that act differently. This can be documented by simulation modelling. We have done a lot of this work. In one of the farms we are studying we did a soil survey for 120 ha, and we looked at the variation. The cost was 20 000 guilders. This is a substantial sum, but it is a one-time investment and is acceptable to the farmer if we can demonstrate the benefit. The step from the 'pedological' identity to the 'functional' identity is complicated, but as we do this we build up a database. I foresee that in the near future we will be able to go to a farm, do a soil map, and make 'quick and dirty' assumptions based on the assessment made elsewhere for identical soil series. This will be a new injection for field pedology, building up databases for individual soil series.

Groffman: Establishing relationships between soil type and water and nutrient behaviour has been done for many years. I remember reading papers on this from 1986. This could be a strong approach when combined with precision management.

Mulla: The data Dick Webster showed give us reason to think that precision farming is going to be something that can be accomplished in the field. The first thing that I noticed was that the size of those regions which have different properties are large enough to identify fairly easily and to manage. They are on a scale of about 100 m. Second, there was a relationship between one variable and another and we could make use of this to simplify the sampling strategies. For instance, if a general relationship with yield exists, we could monitor the yield with remote sensing or with yield monitors and we could have a targeted sampling strategy where we sample more intensively in the areas where yield is changing quickly, so we don't have to take a sample every 50 m in order to identify those zones where the management has to be changed. I appreciate the data you have shown, and I think they tell us something about how we can really simplify the practice of precision farming.

Rabbinge: Many of you today have emphasized that it is important clearly to define the background from which you are working and what the purpose of the activity is. When you are trying to improve the management of a farming system or when you're over-using inputs considerably, your aims are different. In all cases, however, fine-tuning is the appropriate instrument for achieving better productivity, increased efficiency and efficacy as the ultimate goals. Here in Wageningen we are just 40 km from a small town called Bergh, which holds a record in The Netherlands, which they are not very proud of, for the highest surplus of nitrogen per hectare—700 kg N/ha per year—due to the intensive cattle breeding that was going on and efforts to get rid of dung. If you want to reduce that to 200 kg N/ha, this is a big step, but the level is still much too high. So I'd say that the background is saying something about where you are going.

Predicting wheat yields: the search for valid and precise models

V. Barnett, S. Landau†, J. J. Colls†, J. Craigon†, R. A. C. Mitchell* and R. W. Payne*

*Department of Mathematics, University of Nottingham, University Park, Nottingham NG7 2RD, †Department of Physiology and Environmental Science, University of Nottingham, Sutton Bonnington Campus, Sutton Bonnington, Loughborough, Leicestershire LE12 5RD and *IACR-Rothamsted, Harpenden, Hertfordshire AL5 2JQ, UK*

Abstract. Interest in predicting wheat yield in terms of physiological, cultural and meteorological variables is more than a century old. Early attempts involved statistical analyses of relationships between yield and observational data on precipitation, temperature, radiation, etc., and scientific study of physiological and cultural influences such as dates of sowing or anthesis, farming procedures and soil treatments. More recently these have been augmented by large-scale mechanistic models of phenological development, such as AFRCWHEAT, CERES and SIRIUS, incorporating some simulation facilities. All approaches implicitly involve fitting models of some sort: statistical, mechanistic or (preferably) a hybrid of these forms. Levels of success on this important matter are highly variable. After reviewing the field, we consider the results of recent efforts to contrast and evaluate the (large-scale) mechanistic approaches, using spatial/temporal methods for interpolating the required climatological input variables. The work employs a substantial database of wheat yields assembled for this purpose. After assessing the validity of the large-scale mechanistic models (with some intriguing conclusions), we then consider some results from a current approach to parsimonious hybrid modelling, based on statistical study of accessible climatological data interpreted in terms of physiological knowledge of key influences on plant development.

1997 Precision agriculture: spatial and temporal variability of environmental quality. Wiley, Chichester Ciba Foundation Symposium 210) p 79–99

An interest in predicting crop growth in the field is a natural part of our concern to monitor agricultural, financial and development matters. The interest arises at all levels: farmer, economist, scientist and politician. At the micro scale, the farmer needs to predict the likely returns from different crop choices and farming practices. At the macro scale, governments and communities of nations need to plan and plot their agricultural policies. Whilst the farmer and the politician might watch with concern what happens in the field, they must be equally concerned with what the scientist is discovering in the laboratory, or in field-plot experiments, or in the construction of formal models for crop growth utilizing mechanistic (scientific) and statistical (experiential) components. This general crop interest carries over to specific crops, the most commonly considered of which is winter wheat, the topic of this review.

The early days

Observation, monitoring and, by implication (and later by design), the prediction of winter wheat growth goes back a long way. In Rothamsted (the earliest agricultural research station in the world, established in 1843), its founder John Bennet Lawes from the outset sought to understand how climate and the new chemical fertilizers he was developing were affecting crop growth, including growth of wheat. With great foresight (see Barnett 1993, 1994) he set out the renowned large-scale 'classical' experiments, on pioneering quasi-statistical principles, which have been continued to the present time providing a veritable environmental time warp of materials (seeds, soil, etc.) and tabulated measurement information throughout the last 150 years.

The 'Broadbalk' experiment was laid down to wheat in 1843 and has remained virtually unchanged in its form to the present day. Full climate data and yields were monitored throughout, illustrating Lawes' conviction of the important effects of rainfall and temperature on crop yields (Lawes & Gilbert 1871). The vast range of materials and data from the Rothamsted classical experiments have been catalogued and mounted in computerized database form as the Electronic Rothamsted Archive (Potts et al 1996). With the appointment of R. A. Fisher in 1921, formal statistical methods (in inference, regression and designed experiments, reflecting the crossed-factor style of the layout of the 'classical' experiments) were rapidly devised to analyse the relationship between crop development and climate. Fisher (1921) exhibited, by use of orthogonal polynomials, strange residual cyclic patterns in wheat growth on Broadbalk over its first 70 years, after allowing for the 'annual effects' of weather (including rainfall). (See also Fisher 1924.)

Empirically based statistical studies (using regression and correlation methods), often featuring growth of wheat along with other crops, continued throughout this century, for example in the work of Hooker (1907, on effects of residual temperatures above 42 °F), Wishart & Mackenzie (1930, rainfall and barley), Cashen (1947, rainfall and grass), Smith (1960), Buck (1961, including temperature and transpiration), Jenkinson et al (1994) and Chmielewski & Potts (1995, precipitation and minimum and maximum air temperature effects on grain yield on the single site of Broadbalk).

Statistical and mechanistic methods and models

The work described above had two characteristics. It examined *data-based empirical (statistical) links*, principally by means of regression or correlational models with at most informal regard for plant-physiological effects, and it was *restricted to single-site studies* often under strictly controlled experimental conditions. Degrees of relationship were found between yield and climatological variables with multiple correlation coefficients often in the range 0.5–0.6, thus explaining about 25–35% of the variability. Obvious related questions arose and stimulated research from different standpoints. What could we say about:

(1) the scientific influence of climate and chemical substances during phenological development stages of crop (wheat) growth; and
(2) the possibility of incorporating such knowledge in complex mechanistic models to predict yield in terms of all available information over the whole growth period?

Such efforts moved the emphasis away from statistical study of empirical/observational links between weather and yield to examining the contributing biological processes under controlled conditions. There is extensive understanding of the effects of environment on the physiological processes determining yield (see e.g. the reviews by Evans 1993 and Fischer 1983).

In the second category, highly specific developments have taken place where attempts have been made to formulate this knowledge into models. Three major wheat models appeared: AFRCWHEAT (see Porter 1993, for effectively its latest form), CERES-wheat (Ritchie & Otter 1985) and SIRIUS (Jamieson et al 1997) were developed in the UK, the USA and New Zealand, respectively.

Detailed (day-by-day) climate effects are central to these models. Many other factors, such as water and nitrogen effects, also feature. Nonhebel (1993) compared model predictions (for the SUCROS model) from observed and interpolated weather inputs, using actual weather data over (just) 33 years from one site, namely Wageningen; Aggarwal (1995) considered model sensitivity to uncertainties (due to measurement, calibration, etc.) in the crop, soil and weather inputs, contrasting simulation with the alternative sensitivity analysis approach.

The use of large-scale models is typified in the recent paper by Rosenzweig & Tubiello (1996), who begin by observing that recent observations and general circulation models (mechanistic models of climate) predict that a feature of global warming will be marked asymmetry between daytime maximum temperatures and night time minima. CERES-wheat is then used to predict wheat yields in the central USA by running simulations at four sites under various climate-change scenarios. In this work we find almost no observational reference base, just four sites and one large-scale mechanistic model driving another with outcomes expressed through simulations.

Thus returning specifically to the prediction of wheat yields, we have three powerful models, widely applied to examine wheat growth. While until recently relatively little effort had been made to validate these models using substantial real-life data sets on yields and actual climate (and other) input information, or to make major inter-site comparisons, there are signs that interest in this matter is growing. Otter-Nacke et al (1986) is perhaps the most substantial study to date, for CERES-wheat. (See also Porter et al 1993, Wolf et al 1995.) A major thrust is evident in the work of GCTE (see Anon

1994). On the other side of the coin, attempts to develop sophisticated multi-site empirical (statistical) models have also been lacking.

Wheat model validation

Thus obvious avenues of enquiry are suggested by the following questions:

(1) Can we assemble an extensive multi-site real-data set of yields, climate and other inputs to examine how well the models (AFRCWHEAT2, CERES-wheat and SIRIUS) work in practice (i.e. to seek to validate them)?
(2) Can we develop a parsimonious empirically based statistical model (with relatively few parameters, easily applied) incorporating growth-development information, as an alternative to such models, and how well does it behave in the broader multi-site environment?

The UK Biotechnology and Biological Science Research Council has funded a collaborative research study (involving IACR-Rothamsted and Nottingham University) on precisely these lines, and in the rest of this paper we outline some of the preliminary findings of this work. It is conveniently divided into various stages.

The wheat yield database

As raw material for the validation and modelling components, there was need for an extensive electronic database of wheat yields with detailed related phenological and climatological information. For validation of the wheat models the climate data were needed at the daily level! No such combined data set was in existence, so it was necessary to compile one from results throughout the UK and over many years of well-managed wheat trials. (Hudson [1996] used a more limited set of data to study nitrogen response and a MAFF-funded project in the Statistics Department at Rothamsted provided an initial boost to the data-assembly process.) This involved extensive effort, which yielded the appropriate level of detail for more than 1000 wheat trials. Data covered an 18 year period (from 1975) and yields were augmented with information on crop treatments, trial design, cultivar, sowing date and grid reference and altitude. Of course, the immediately relevant climatological data were not included since detailed meteorological records were unlikely to have been collected at the trial sites. Having thus assembled the crop database, the first research task was to seek to provide the relevant associated climate data.

Climate data interpolation

The trial sites were distributed over the UK and their grid references and altitudes were known. Two approaches were possible. We could either seek to model the weather fluctuations, or use real-life weather information. Some interesting mechanistic models have been developed (e.g. Hutchinson 1995). Indeed, such methods have been introduced in the wheat models themselves in recent applications to explore global warming effects. However, they were not appropriate for the present exercise of attempting to validate the wheat models for real data, where clearly we need to examine actual yields and weather conditions.

Thus the weather data for input into the models had to be found from a source other than the compiled wheat database. Specifically, we needed temperature, rainfall and radiation (or sunshine hours) on a daily basis. These were determined from the Meteorological Office database, covering the set of 212 meteorological stations, maintained at IACR-Rothamsted. Since the locations of these stations do not coincide with those of the wheat trials it was necessary to develop reliable interpolation methods to estimate the weather conditions encountered at the trial sites (see Landau & Barnett 1996). A subset of about 1/3 of the trial sites is shown in Fig. 1.

Climate variable interpolation is a matter of much scientific and commercial interest. However, previous (published) work (e.g. Hulme et al 1995, Lennon & Turner 1995) was not disaggregated down to the level of daily detail, and no detailed methods have been proposed that co-ordinate the spatial and temporal dimensions of variation (but see Haslett 1989, Hancock & Wallis 1994 and Mardia & Goodall 1993 for discussion of spatial/temporal principles in weather modelling).

Landau & Barnett (1996) describe four, increasingly sophisticated, methods for interpolating minimum and maximum temperatures, sunshine duration and accumulated rainfall each day over the growing season at each of the trial sites. It was clear that spatial and temporal variations had to be accounted for and this was achieved by a judicious combination of kriging, regression and time series methods.

Recalling that the grid-point references and altitudes of all meteorological stations and trials sites were known, it was feasible to contemplate three spatial regressor variables; the continuity of meteorological records over decades made it possible to consider fitting time-series models to reflect seasonalities and time-local associative patterns. Fuller details are given by Landau & Barnett (1996), but in summary the four methods progressed through the following stages:

(1) Direct (deterministic) interpolation was carried out in the form of a weighted combination of adjacent observations with combined spatial/temporal weight function.
(2) A more global statistical view was incorporated by regressing on latitude, longitude and elevation for each day separately.
(3) Trend surface residuals were then kriged to reintroduce a 'random' daily perturbation of the trend surface values.
(4) Time-variational components were introduced by fitting time-series models over 7 years at each meteorological station, followed by spatial regression of three relevant parameters in the time-series models and then by kriging of the residuals.

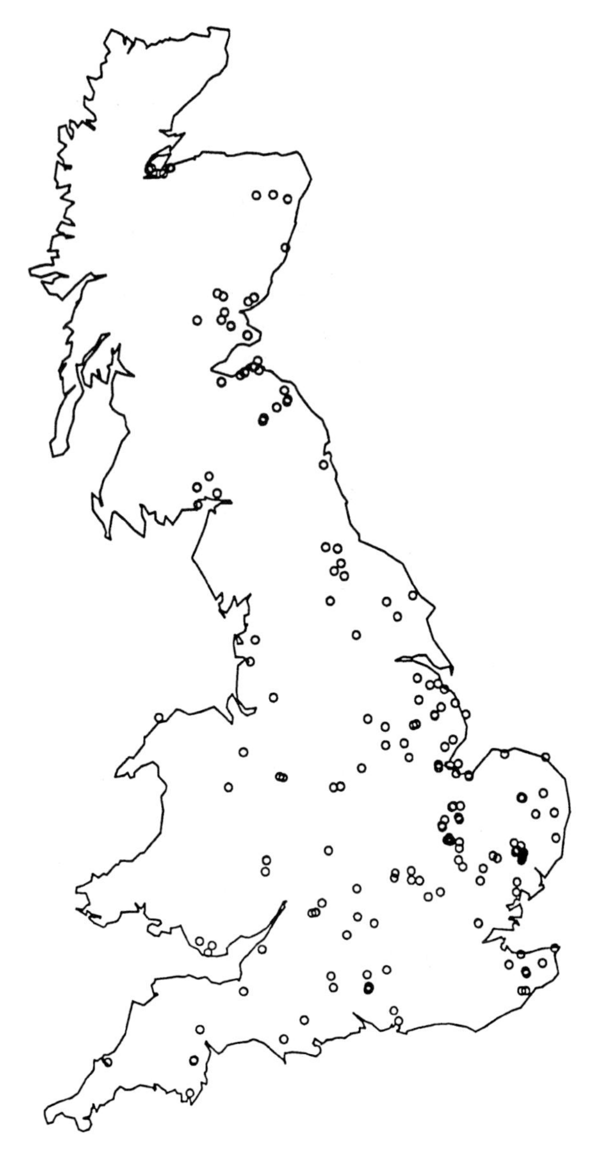

FIG. 1. Wheat trial sites in the UK. (From Landau et al 1997.)

How did this all work out? Validation was carried out on about 10% of the locations and times, predicting from each of the methods the values of the climatological variables for comparison with the true values. Mean squared errors were used to compare the methods. No large differences were found. The primitive first method

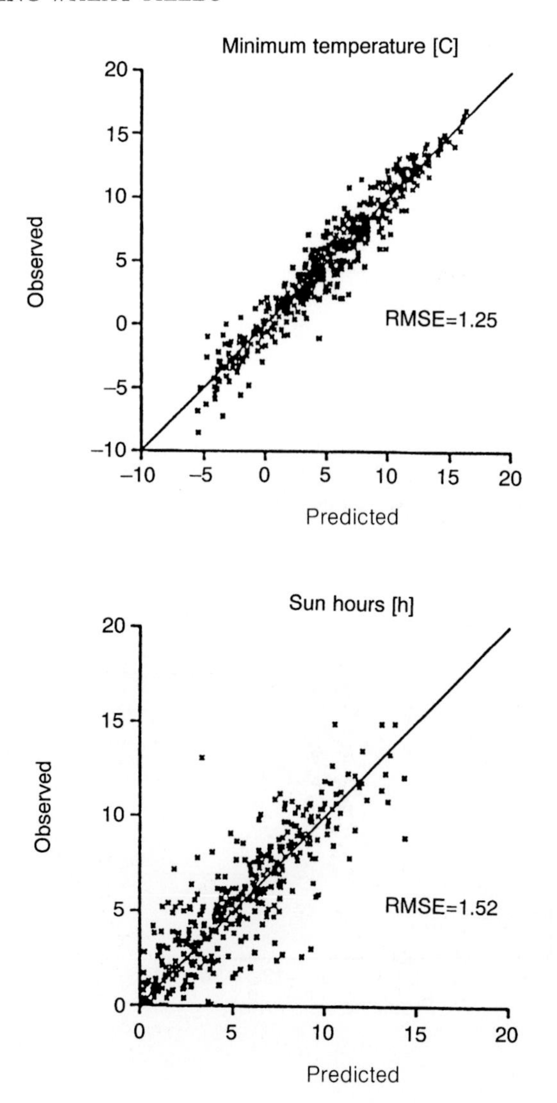

FIG. 2. Observed and predicted minimum temperature and sunshine hours in 1987 for the trial sites in Fig. 1 (see text for details). (From Landau & Barnett 1996.)

of direct interpolation achieved results impressively close to those of the more sophisticated ones. Overall, the third and fourth approaches did best and in view of the more parsimonious parameterization of the spatial/temporal fourth stage approach, this was chosen as the climate interpolator for the later stages of the work. Figure 2 compares the observed and predicted values of minimum temperature and

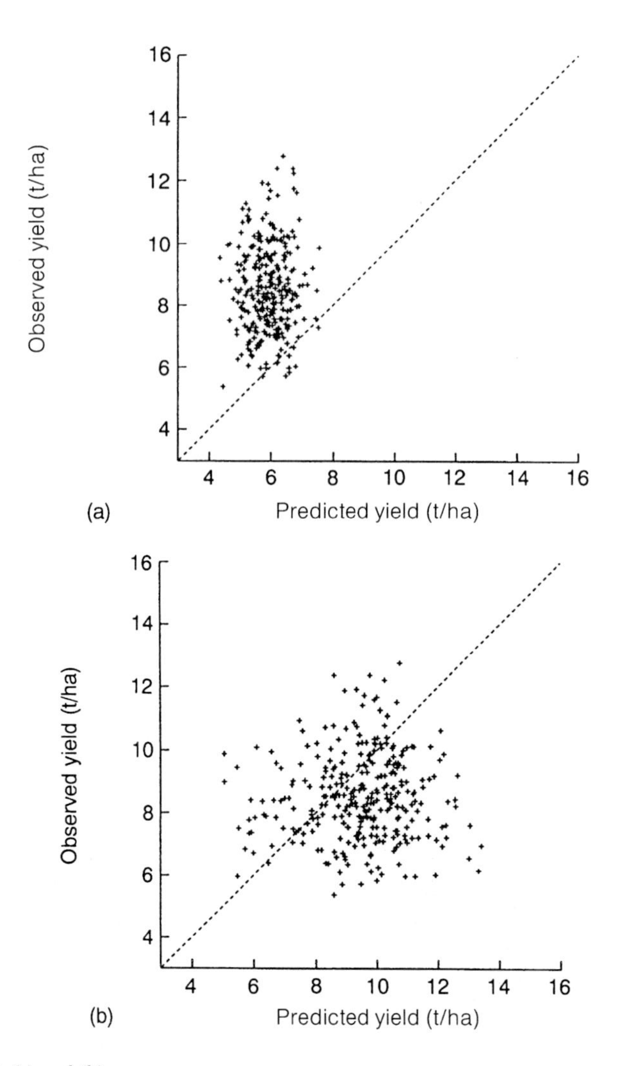

FIG. 3. (a) and (b).

sunshine hours for 1987 using this approach. Sunshine hours and rainfall were more difficult to predict than temperatures, for obvious reasons to do with the distributional forms of the variables and their 'patchiness' of coverage in the meteorological database. None the less, approaching 85% and 75% of the variations, respectively, were explained by the fit, compared with almost 95% for temperatures.

Daily solar radiation is a vital input to the crop models. This was approximated using standard methods from interpolated sunshine duration taking the seasonal effects into account.

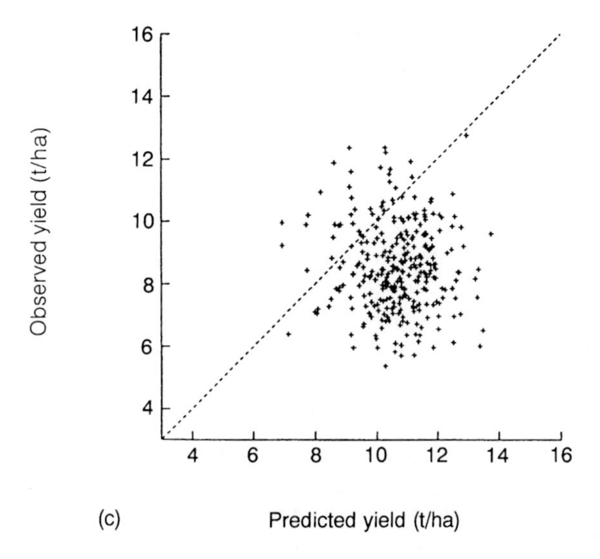

FIG. 3. Observed and predicted wheat yields. (a) CERES-wheat; (b) AFRCWHEAT2; and (c) SIRIUS. (Modified from Landau et al 1997.)

Thus we were now armed with an extensive database of wheat yields, trial characteristics, and (interpolated) daily weather details with which to embark on an attempted validation of the three wheat models.

The empirical performance of AFRCWHEAT2, CERES-wheat and SIRIUS

An initial study (sensitivity of predicted yield and anthesis date to interpolation errors in the climate variables) is described in Landau & Barnett (1996). More detailed validation studies have now been completed (see Landau et al 1997). The conclusion (taken from the abstract) is that 'none of the models accurately predicted historical grain yields between 1976 and 1993. Substantial disagreement was found between the models' predictions . . .', which is at first sight most damaging to the claims of these models to predict grain yields of wheat. Figure 3 shows observed and predicted yields. We can clearly see the lack of predictive link for each of the models, with additional major biases for CERES-wheat (Fig. 3a) and SIRIUS (Fig. 3c). In no case was there *any significant correlation* between observed and predicted yield. The models also did not markedly mirror each other's predictions, with estimated inter-correlation coefficients of about 0.24 between AFRCWHEAT2 and each of the others, and of about 0.35 between CERES-wheat and SIRIUS.

So what are we to conclude? Are the models seriously flawed? Is there some basic fault in this extensive validation study? Landau et al (1997) discuss these matters. It seems that one of the main reasons is that weather effects on wheat yields in the UK tend to be related to agronomic factors such as the ability to harvest or to control diseases, rather then physiological ones. The wheat models are designed to be applied under ideal management so that they cannot, in their present forms, predict variations in yields under conditions which exist in practice in the UK.

A new parsimonious hybrid model

The above conclusions that statistical models are often single-site based and do not consider phenological or scientific effects on crop yield and that large-scale mechanistic models, whilst multi-site oriented, do not seem to have any predictive value, pose a challenge in an obvious direction. Could we formulate a parsimonious statistical model which is applicable across a range of sites, which incorporates phenological growth-stage factors and which yields a useful level of predictive power (e.g. multiple correlation of the order of 0.5–0.6)?

The wheat database has been extensively interrogated specifically with this aim, and early results are indeed encouraging.

As in previous (single-site) studies, certain broad effects were clearly distinguishable. All three climatological measures (temperature, rainfall and radiation) are influential, but particularly when considered on a selective basis in relation to the growth period and the different growth stages. Whilst some effects might be superficially surprising, they become interpretable against the scientific knowledge of effects on growth and development.

Growth characteristics interrelate with climatological measures differently at different stages of the growth cycle (Landau 1997). Radiation supports the process of photosynthesis and drives growth; it supplies the assimilate for grain-filling. Temperature is clearly a major influence, especially as expressed through 'thermal time', but has a threshold effect at high levels. It is important as it shortens duration of grain fill, and can advance senescence of leaves to the detriment of yield. High temperatures in the month before anthesis can be disadvantageous. Severe winter frost can cause serious damage. Inadequate rainfall, particularly at anthesis, can reduce yield. The various climatological influences can also interact subtly with each other; soil-type is also relevant to such interactions. Indirect weather effects are also important, e.g. for pest, fungus or weed stimulus or reduction of nitrogen (although these factors are not included in the mechanistic wheat models).

Such relationships were considered in exploratory studies of the empirical links between climate and yield, based on dividing the growth period into five phases (vegetative, early reproductive, anthesis, grain-filling and pre-harvest) and examining important factors at the different stages. This leads to the following initial choices for major climate influences in order of importance:

(1) rainfall from early reproductive to grain-filling;
(2) radiation from early reproductive to grain-filling;
(3) extreme temperatures (particularly frosts in January);
(4) indirect climate effects at drilling and harvest; and
(5) other climate effects during the vegetative period.

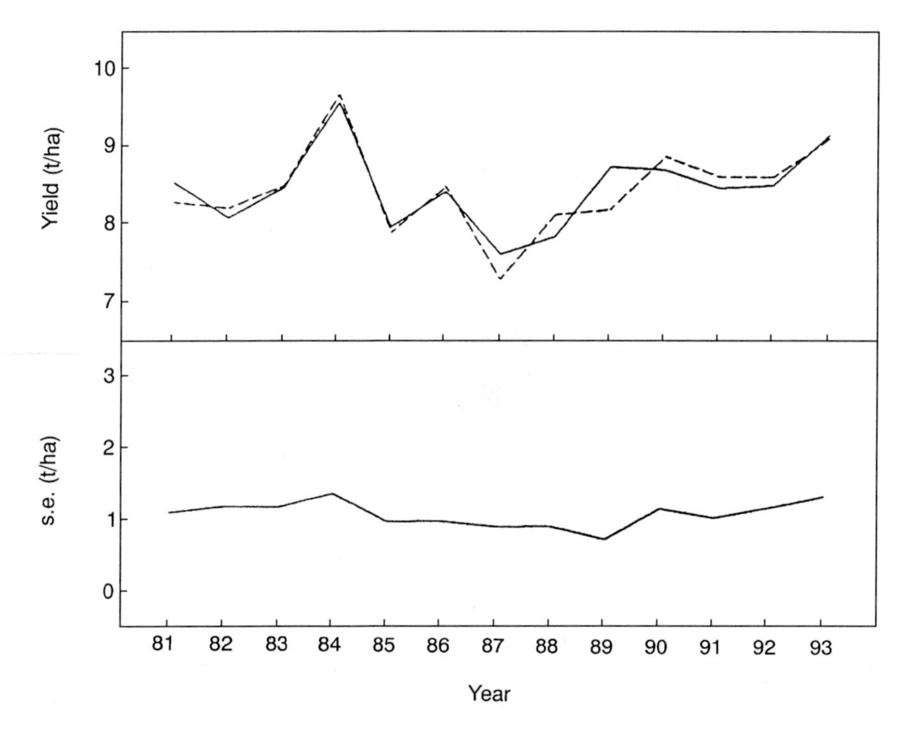

FIG. 4. Observed (broken line) and predicted (solid line) annual mean wheat yields, with standard errors (see text for details). (From Landau 1997.)

Problems of multicollinearity needed special attention.

The way forward was to choose a set of explanatory (physiologically oriented) climate variables as the 'stock' from which the final drive variables of a parsimonious phenological/empirical model would be chosen. If various models are of similar predictive value, the final choice would reflect qualitative features based on physiological expectations and distinct empirical features.

A very detailed investigation of alternative regression models was carried out using more than 30 stock variables, and with elimination or retention of variables (on various statistical, physiological and agronomic criteria) in the search for models which:

(1) had good predictive value;
(2) had only a modest number of parameters; and
(3) accommodated identified physiological/climatological relationships.

The models applied of course on a multi-site basis and were ground-breaking not only in this respect, but also in the major characteristics (1), (2) and (3). Space does not allow a detailed description of the route taken to the final models, nor even of their general, let alone explicit, forms. What is remarkable is that a similar order of

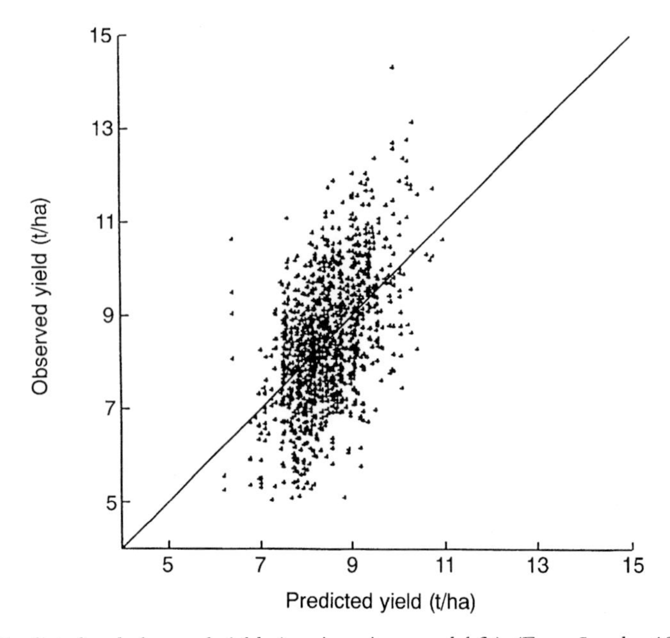

FIG. 5. Predicted and observed yields (parsimonious model fit). (From Landau 1997.)

multiple correlation (0.5–0.6) was obtained as applied in the earlier restricted single-site studies, that the models employed agronomic and physiological as well as climate variables, and that they produced no major bias in the predicted yield. Figures 4 and 5 give some indication of how well the models work. In all such respects they clearly advance our knowledge of how to model wheat yield. They may also prove useful in refining the simulation wheat models described above to improve their predictive capabilities.

References

Aggarwal PK 1995 Uncertainties in crop, soil and weather inputs used in growth models: implications for simulated output and their applications. Agric Syst 48:361–384

Anon 1994 International Geosphere Biosphere News. 18 June 1994

Barnett V 1993 Multivariate environmental statistics in agriculture. In: Patil GP, Rao CR (eds) Multivariate environmental statistics. Elsevier, Amsterdam, p 1–32

Barnett V 1994 Statistics and the long-term experiments. In: Leigh RA, Johnson AE (eds) Long-term experiments in agriculture and ecological sciences. CAB International, Wallingford, p 165–183

Buck SF 1961 The use of rainfall, temperature, and actual transpiration in some crop–weather investigations. J Agric Sci 57:355–365

Cashen RO 1947 The influence of rainfall on the yield and botanical composition of permanent grass at Rothamsted. J Agric Sci 37:1–9

Chmielewski F-M, Potts JM 1995 The relationship between crop yields from an experiment in Southern England and long-term climatic variations. Agric Forest Meteorol 73:43–66

Evans LT 1993 Crop evolution and yield. Cambridge University Press, Cambridge

Fischer RA 1983 Wheat. In: Potential productivity of field crops. Proceedings of a Symposium at IRRI, Manila, Phillipines, p 129–154

Fisher RA 1921 Studies in crop variation. I. An examination of the yield of dressed grain from Broadbalk. J Agric Sci 11:107–135

Fisher RA 1924 The influence of rainfall on the yield of wheat at Rothamsted. Philos Trans R Soc Lond Ser B Biol Sci 213:89–142

Hancock S, Wallis R 1994 An approach to statistical spatial-temporal modeling of meteorological fields. J Am Stat Assoc 89:368–378

Haslett J 1989 Space time modelling in meteorology. Bull Int Stat Inst 51:229–246

Hooker RH 1907 Correlation of the weather and crops. J R Stat Soc 70:1–42

Hudson D 1996 Analysis of the effects of treatments in non-linear models for nitrogen response, with implications for design. PhD thesis, University of Cambridge, Cambridge, UK

Hulme M, Conway D, Jones PD, Jiang PD, Barrow EM, Turney C 1995 Construction of a 1961–1990 European climatology for climate change modelling and impact applications. Int J Climatol 15:1333–1363

Hutchinson MF 1995 Stochastic space–time weather models from ground-based data. Agric Forest Meteorol 73:237–264

Jamieson PD, Semenov MA, Brooking IR, Francis GS 1997 SIRIUS: a mechanistic model of wheat response to environmental variation. Field Crops Res, in press

Jenkinson DS, Potts JM, Perry JN, Barnett V, Coleman K, Johnston AE 1994 Trends in herbage yields over the last century on the Rothamsted Longterm Continuous Hay Experiment. J Agric Sci 122:365–374

Landau S 1997 A parsimonious model for yield response to environment. PhD thesis, University of Nottingham, Nottingham, UK

Landau S, Barnett V 1996 A comparison of methods for climate data interpolation, in the context of yield predictions from winter wheat simulation models. In: White EM, Benjamin LR, Brain P et al (eds) Aspects of applied biology, modelling in applied biology: spatial aspects. Asoociation of Applied Biologists, Wellesbourne, UK, p 13–22

Landau S, Mitchell RAC, Barnett V et al 1997 Testing wheat simulation models' predictions against an extensive data set of observed grain yields in the UK. Agric Forest Meteorol, submitted

Lawes JB, Gilbert JH 1871 Effects of the drought of 1870 on some experimental crops at Rothamsted. J R Agric Soc 7:91–132

Lennon JJ, Turner JRC 1995 Predicting the spatial distribution of climate: temperature in Great Britain. J Anim Ecol 64:370–392

Mardia KV, Goodall CR 1993 Spatial-temporal analysis of multivariate environmental monitoring data. In: Patil GC, Rao CR (eds) Multivariate environmental statistics. Elsevier, Amsterdam, p 347–386

Nonhebel S 1993 The importance of weather data in crop growth simulation models and assessment of climate change effects. PhD thesis, Agricultural University of Wageningen, The Netherlands

Otter-Nacke S, Godwin DC, Ritchie JT 1986 Testing and validating the CERES-wheat model in diverse environments. AgRISTARS Technical Report, YM-15-00407, JSC20244. Johnson Space Center Houston, TX

Porter JR 1993 AFRCWHEAT2: a model of the growth and development of wheat incorporating responses to water and nitrogen. Eur J Agron 2:69–82

Porter JR, Jamieson PD, Wilson DR 1993 Comparison of the wheat simulation models AFRCWHEAT2, CERES-wheat and SWHEAT for non-limiting conditions of crop growth. Field Crops Res 33:131–157

Potts JM, Verrier PJ, Payne RW 1996 The electronic Rothamsted archive and its use in agricultural research. Proceedings of 6th International Congress for Computer Technology in Agriculture, ICCTA '96, Wageningen, 16–19 June 1996

Ritchie JT, Otter S 1985 Description and performance of CERES-wheat: a user orientated wheat-yield model. Rep US Dept Agric, ARS 38:159–175

Rosenzweigh C, Tubiello FN 1996 Effects of changes in minimum and maximum temperature on wheat yields in the central US — a simulation study. J Agric Forest Meteorol 80:215–230

Smith LP 1960 The relationship between weather and meadow-hay yields in England. J Br Grassland Soc 15:203–208

Wishart J, Mackenzie WA 1930 Studies in crop variation. VII. The influence of rainfall on the yield of barley at Rothamsted. J Agric Sci 20:417–439

Wolf J, Semenov MA, Eckerstein H, Evans LG, Iglesias A, Porter JR 1995 Effects on winter wheat: a comparison of five models. (G2LU38). In: Harrison PA, Butterfield RE, Downing TE (eds) Climate change and agriculture in Europe. Environmental Change Unit, Oxford, p 231–280

DISCUSSION

Webster: To follow-up from your talk, I have two comments. First, in a sense plant breeders have already tried to take account of weather fluctuations by attempting, with great success, to breed varieties resistant to these fluctuations. If the varieties they develop are not resistant, then they are not commercially successful and they are not marketed. To some extent it is not surprising that you didn't find much relationship between weather and crop yield because of this.

Second, as far as I understand it, you were not able to take account of the variations in soil from experiment to experiment. I think that is something that could and should be done in any future investigations. I am reminded of the experience of Rothamsted and the ICI company in the late 1970s when ICI set up what it called the '10 tonne club' to stimulate farmers to increase their wheat yields. The yields were measured, and any farmer who produced 10 tonnes of grain per hectare over a whole field was admitted to the 'club'. The experience was interesting. Out of around 1000 farmers who applied, few achieved the 10 t/ha, and when Weir et al (1984) analysed the data they discovered that soil type was the single biggest factor influencing yields, from an average on the worst soil type of about 6 t/ha to an average on the best of more than 9 t/ha. This reinforces how important soil type is.

Barnett: With respect to the climate effect argument, the models I looked at described themselves as being designed to model wheat growth under best management conditions. They are constantly honed to relate (according to their developers) to the varieties which are currently on the market for that purpose. They none the less still claim that yield is highly dependent upon climate — so much so that they require four climate variables within the models day by day. This is clearly seen as absolutely integral. So, even if climate is now more extreme or less extreme than it used to be, I

TABLE 1 (*Rasch*) Dependence of leaf area (y_i) in palms on age (x_i)

x_i *(years)*	y_i *(m^2)*
1	2.02
2	3.62
3	5.71
4	7.13
5	8.33
6	8.29
7	9.81
8	11.30
9	12.18
10	12.67
11	10.62
12	12.01

do not accept that this is irrelevant to these models, because in the eyes of the developers it is an absolutely crucial component.

On your second point I agree with you entirely: soil type is a vital determinant. We have attempted to analyse this. We were not able, in the 1000-site wheat database, to get as much information as we would have liked on the soil. Where we have had that information we have done sub-studies, but due to limited coverage these have had to be more-or-less based on looking at certain circumstances for a particular type of soil. We find in all of these sub-studies exactly the same lack of relationship. So I do not know quite what the answer is. If you're saying (and I would have some sympathy with this) that perhaps we should throw everything away and just use soil type as the only determinant, you might indeed do as well. After all, you cannot do worse than these models, but you might do as well as those climate-driven regression models.

Rasch: I think we should return to the problem discussed by Pascal Monestiez yesterday. In the growth curve analysis in plants and animals we switch over from the problem of model construction to the problem of model selection, because only a couple of functions (from which we have to choose the best one) are needed to describe the process you have in mind. The models are parsimonious (they use as few variables as possible) and they are also physiologically interpretative. Perhaps a solution might be to combine the growth curve analysis with the multisite aspects. With some luck we may find a function which is useful for describing the temporal relationship and we can then use this function for different sites. This means that we do not have to change the function but only the parameters from site to site.

I will show you what I mean in the following example. This is the problem of describing the dependence of the leaf area of oil palms on time. The data we used are taken from Rasch (1995) and shown in Table 1 (*Rasch*). They were collected at the Bah Lias Research Station in Indonesia, and describe the yearly growth of the leaf area of oil palms. All calculations are done using the module complex 'growth curve analysis' of CADEMO (see Rasch 1990, Rasch et al 1992).

CADEMO offers a model choice procedure between functions of a subset to be defined from the set of functions to be given in Fig. 1 (*Rasch*). We select the complete set of all 10 functions and use a modification of the Akaike criterion for the model choice (from four criteria available in CADEMO). With the improved Akaike criterion (cf. Hurvich & Tsai 1989), that function is selected as the best for which the expression

$$AIC = n\ln S^2 + n(n+p)/(n-p-2) \quad (1)$$

is minimized ($S^2 = (n-p)s^2/n$, s^2 is the residual variance, n is the number of data used for model selection, p is the number of parameters). This criterion contains a penalty function for the number p of parameters. This means that a four-parametric function in Fig. 1 (*Rasch*) must have a smaller residual variance than a three-parametric one to be better evaluated.

The result of the computation of the AIC values is shown in Fig. 2 (*Rasch*). It can easily be seen that the Bertalanffy function fits best. The fitted function is

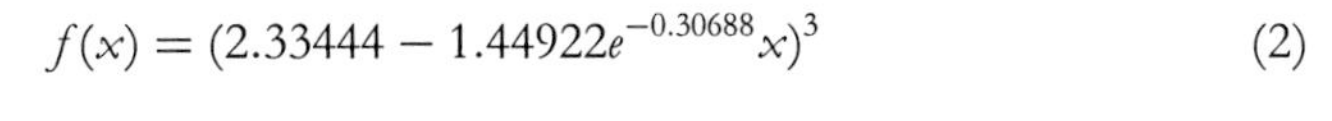

$$f(x) = (2.33444 - 1.44922e^{-0.30688}x)^3 \quad (2)$$

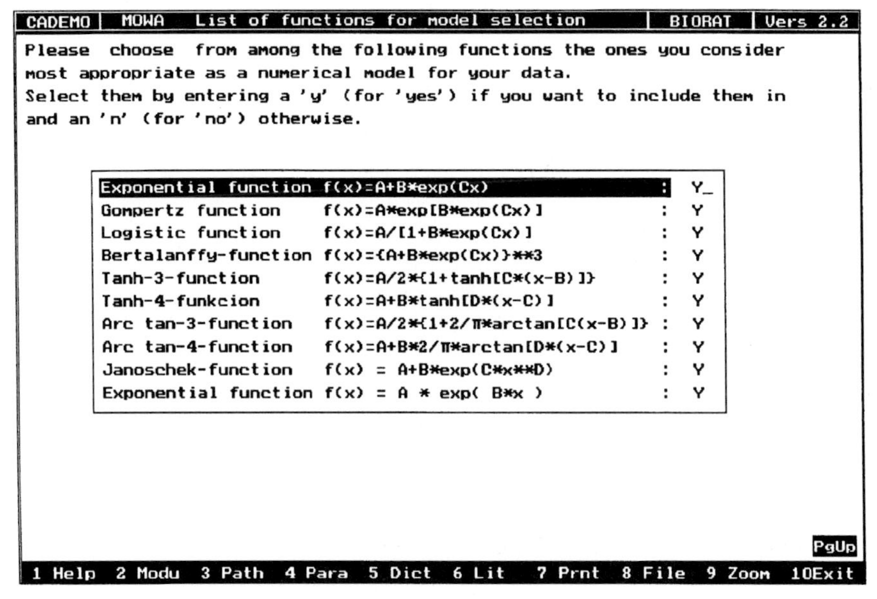

FIG. 1. (*Rasch*) Choice of the set of functions.

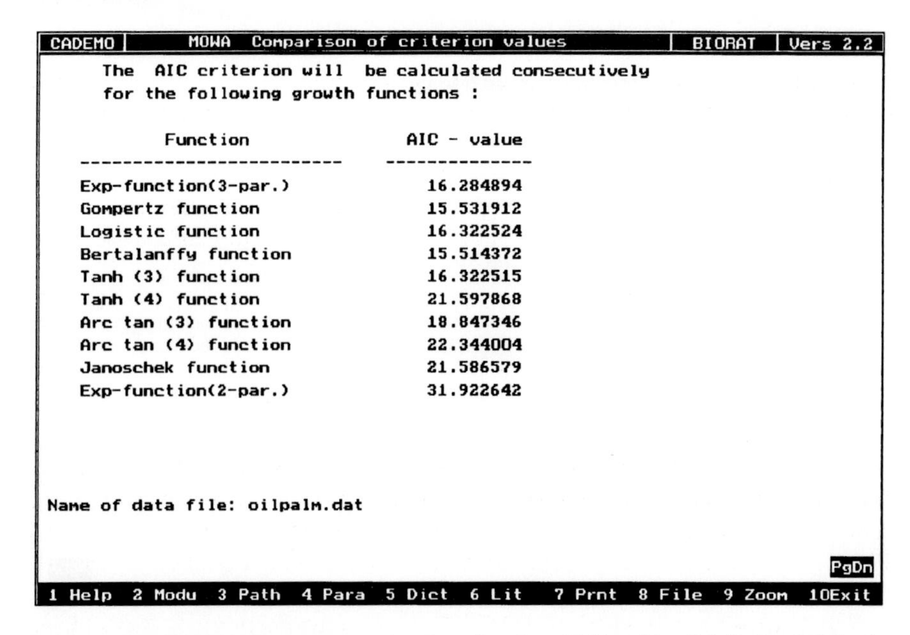
CADEMO | MOWA Comparison of criterion values | BIORAT | Vers 2.2

The AIC criterion will be calculated consecutively
for the following growth functions :

Function	AIC - value
Exp-function(3-par.)	16.284894
Gompertz function	15.531912
Logistic function	16.322524
Bertalanffy function	15.514372
Tanh (3) function	16.322515
Tanh (4) function	21.597868
Arc tan (3) function	18.847346
Arc tan (4) function	22.344004
Janoschek function	21.586579
Exp-function(2-par.)	31.922642

Name of data file: oilpalm.dat

PgDn

1 Help 2 Modu 3 Path 4 Para 5 Dict 6 Lit 7 Prnt 8 File 9 Zoom 10Exit

FIG. 2. (*Rasch*) AIC values from (1) using the values from Table 1 (*Rasch*) (oilpalm.dat) for the 10 functions in Fig. 1 (*Rasch*).

Thus the estimates of α, β and γ are:

$$\hat{\alpha} = 2.33444$$
$$\hat{\beta} = -1.44922$$
$$\hat{\gamma} = -0.30688$$

A graph of (2) is shown in Fig. 3 (*Rasch*). This function really seems to be an adequate model for the growth of the leaf areas of oil palms. If we neglect the two possible outliers, the observations for $x_6 = 6$ and $x_{11} = 11$, we get the following result in the model selection procedure (Fig. 4. [*Rasch*]). The parameter estimates of the fitted function (1) are now:

$$\hat{\alpha} = 2.36789$$
$$\hat{\beta} = -1.49586$$
$$\hat{\gamma} = -0.30205$$

and are similar to those obtained using the complete dataset.

The next step is to determine what the best design is. Here the best design is to have three measurement points in place of 12 and to measure four times at these points. This

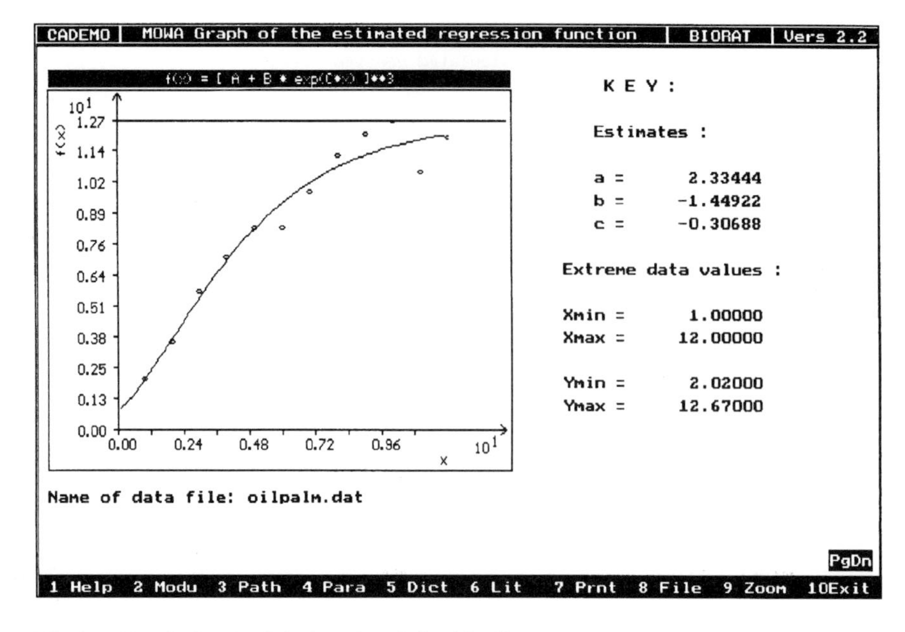

FIG. 3. (*Rasch*) Graph of the function defined in (2).

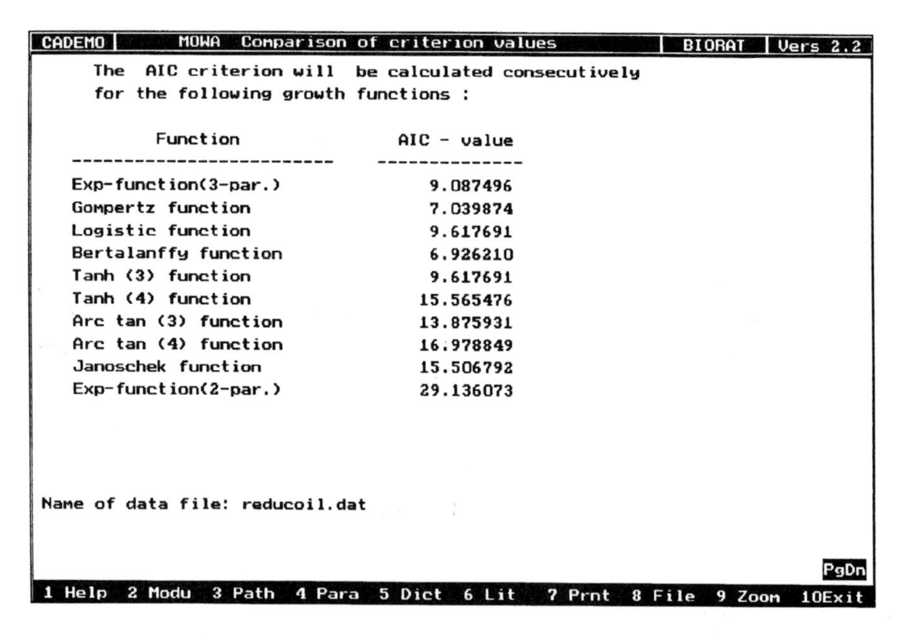
CADEMO | MOWA Comparison of criterion values | BIORAT | Vers 2.2

The AIC criterion will be calculated consecutively
for the following growth functions :

Function	AIC - value
Exp-function(3-par.)	9.087496
Gompertz function	7.039874
Logistic function	9.617691
Bertalanffy function	6.926210
Tanh (3) function	9.617691
Tanh (4) function	15.565476
Arc tan (3) function	13.875931
Arc tan (4) function	16.978849
Janoschek function	15.506792
Exp-function(2-par.)	29.136073

Name of data file: reducoil.dat

PgDn

1 Help 2 Modu 3 Path 4 Para 5 Dict 6 Lit 7 Prnt 8 File 9 Zoom 10Exit

FIG. 4. (*Rasch*) AIC values from (1) using a reduced data set from Table 1 (*Rasch*) (reducoil.dat) for the 10 functions of Fig. 1 (*Rasch*).

is much better than equidistant allocation of the measurements over the area you want to investigate.

Barnett: Parsimony was what drove our whole interest, and we aimed to produce a model which required only a limited range of parameters to reflect different circumstances, different sites and so on. The growth curves, of course, are vitally important, but we have in growth curve analysis in humans and in animals a short-term associative pattern which does not exist in the wheat growth situation. Firstly, if you consider an individual plant, this is not going to explain much about wheat growth over a field, say. Secondly, it is not the growth pattern that is intrinsically interesting, because we are concerned with grain yield: that does not occur until the very end of the process after anthesis. So there seems to be no immediate relevance in the use of growth models.

As far as monitoring is concerned, I am all for it, but we are not going to persuade the commercial powers that be to suddenly place the wheat trials where we want for optimal statistical and physiological monitoring. We have to live with them where they are. I do not think this should be an ongoing process anyway — it is important that we recognize that there are severe limitations in these mechanistic models. I am not persuaded by the argument that 'they explain things, but they do not happen to fit the data': this seems to be a rather strange way of describing what a model does. We know the physiology behind the models. But if it is then put into a model and that model achieves nothing, what is its significance?

Mulla: The models which you described are very data intensive. You can't evaluate their performance over a large area where there's so much uncertainty in the input data that is used to drive them. This point is critical in terms of understanding whether or not the models themselves are valid. There have been a fair number of single-site evaluations of these models and they do perform rather well when there are adequate data to drive them.

Barnett: Sorry, but that is not my experience at all — it is not what I found in my full literature search.

Mulla: Perhaps there is an approach somewhere in between the sophisticated mechanistic modelling and the rather empirical regression modelling. This is to take some knowledge of what plants respond to. For example, there is a relationship between evapotranspiration and grain yield. Perhaps you could divide your datasets into subsets, one for calibration and one for validation. You could take an empirical model of the relation between yield and evapotranspiration and try to determine the crop coefficient that is adequate for one subset of your data and then to use the other set for validation.

Barnett: Yes, we did do that.

McBratney: My point is fairly similar to David Mulla's. Do you think that these mechanistic models are over-parameterized? How would you go about coming up with a model that was mechanistic and not over-parameterized and therefore able to predict? In other words, how might you devise a mechanistic model that is also parsimonious?

Barnett: At every stage in the development of these models, efforts have been made to put in the best scientific knowledge. That knowledge in itself is a model. Frankly, I take the pessimistic view: I see no point in pulling these details together to produce a great conglomeration which does not seem to achieve anything more than any of the individual parts of the understanding. Agriculture is not alone in this effect. If you look at the global circulation models you sometimes see the same thing. I have used mechanistic models successfully in the past but they have usually not been on anything like the grand scale of these wheat yield models. I think we should be using the understanding we have in the interstices, as it were. We tried to do this in our empirical approach by not even entertaining relationships which showed up in the data which could not be supported by proper scientific and physiological understanding, and by looking for relationships where we knew such things should be found. So it is a hybrid approach, but it leads to a somewhat untidy model — not a nice little black box you can carry around with you.

Goense: You said that in these models you couldn't find a relationship between the different soil types.

Barnett: It was only a very small subset of the data where we had adequate soil information. But within those subsets we did not find a relationship, because the models themselves don't discriminate that well in terms of soil type.

van Meirvenne: Just as with spatial predictions, where we use predicted values in combination with the prediction variance, these models predict a yield, so that means that the values you obtain should be considered within the confidence interval. I always feel uncomfortable about comparing observed versus predicted values just on the basis of a single point. I'm saying this not as a criticism of Vic Barnett's work, but as a criticism of the model builders. They should include in their models some error propagation modules and produce predictions with confidence intervals.

Barnett: We did all the appropriate sensitivity-type analyses to which you implicitly refer. First of all, a comparison of a thousand observed and a thousand predicted is a grand confidence statement in itself, but, that apart, we then looked at these predictions over the whole range of values that were reasonable on a predicted basis and used confidence interval criteria. We also looked at all the scatter plots derived over this whole range of prospects. In none of them was there any relationship. We were conscious of the point you made and sought to cover it.

Bregt: I recall a study our institute did for the European Union where we constructed a crop growth monitoring system for estimating various yields. Our experience was that models alone predicted yields poorly. However, we still used the models, not for absolute predictions, but for correcting the trends. You get better predictions by using a combination of, for instance, the normal trend and the model outcome.

Barnett: I have not said these models are useless; I have said that they do not provide any facility for predicting wheat growth. As for the question that year-by-year they might be telling you about the macrovariational elements, there is no indication of this either.

Bregt: I agree.

Barnett: I thought that is what you were saying: that they had that more global value.

Bregt: What I was saying is that they cannot be used to estimate the level of yield, but they can be used to correct the trend which can be expected from the general statistics on wheat yield.

Barnett: No; they did not have any such relative value in them, either.

Rabbinge: I think it's worth underlining that different sorts of predictions may be aimed at, and that this may affect which factors are involved. The scheme we use in the field of production ecology is that we are distinguishing different production levels and different production situations. We distinguish a potential production level, the attainable level and the actual level. The potential level is reached in a situation where only the yield-defining factors such as ambient CO_2, solar radiation, temperature and the crop characteristics play a role. The majority of explanatory models are developed for potential situations, which seldom occur. At least 95% of agriculture takes place at levels which are far below the potential situation. The attainable level is dictated by limiting factors such as water and nutrient supply. Crop growth-reducing factors play a role and may result in the actual production level. It is the mixture of different factors that should be considered. If you are looking in the traditional dose–effect way that is not important, but if you are trying to explain things you have to take into account the yield-defining factors and the yield-limiting and yield-reducing factors. As I have already mentioned, models are mainly developed for potential situations where limiting factors are not playing a role. In order to predict yield in any specific situation all these factors must be taken into account. In a review article (Loomis et al 1979) we showed that the regression type of descriptive models can be used more effectively if they contain some explanatory value than the pure mechanistic explanatory models. This is because if you want to use explanatory models you have to take into account all these different factors.

References

Loomis RS, Rabbinge R, Ng E 1979 Explanatory models in crop physiology. Annu Rev Plant Physiol 30:339–367

Rasch D 1990 Optimal experimental design in non-linear regression. Comm Statist Theory Meth 19:4789–4806

Rasch D 1995 The robustness against parameter variation of exact locally optimum experimental designs and growth models: a case study. Comput Statist Data Anal 20:441–453

Rasch D, Guiard V, Nürnberg G 1992 Statistische Versuchsplanung — Einführung in die Methoden und Anwendung des Dialog — systems CADEMO. G Fischer Verlag, Stuttgart

Hurvich M, Tsai CL 1989 Regression and time series model selection in small samples. Biometrika 76:297–307

Weir AH, Rainer JH, Catt JA, Shipley DG, Hollies JD 1984 Soil factors affecting the yield of winter wheat: analysis of results from ICI surveys 1979–80. J Agric Sci 103:639–649

Geostatistics, remote sensing and precision farming

D. J. Mulla

Department of Soil, Water and Climate, Borlaug Hall, University of Minnesota, 1991 Upper Buford Circle, St. Paul, MN 55108–6028, USA

Abstract. Precision farming is possible today because of advances in farming technology, procedures for mapping and interpolating spatial patterns, and geographic information systems for overlaying and interpreting several soil, landscape and crop attributes. The key component of precision farming is the map showing spatial patterns in field characteristics. Obtaining information for this map is often achieved by soil sampling. This approach, however, can be cost-prohibitive for grain crops. Soil sampling strategies can be simplified by use of auxiliary data provided by satellite or aerial photo imagery. This paper describes geostatistical methods for estimating spatial patterns in soil organic matter, soil test phosphorus and wheat grain yield from a combination of Thematic Mapper imaging and soil sampling.

1997 Precision agriculture: spatial and temporal variability of environmental quality. Wiley, Chichester (Ciba Foundation Symposium 210) p 100–119

Interest in precision farming has exploded in the last five years. Much of this interest has occurred in universities and agricultural industries. There now exist several precision agriculture centres at research universities and faculty at these institutions have organized several major international conferences, symposia and workshops on this topic. In agricultural and technological industries there is great interest in the new markets developing as a result of precision agriculture. Agricultural industries anticipate that a significant portion of their new business in the next 20 years will result from providing new services and technology in precision agriculture. Indeed, there has already been explosive growth in the sales of yield monitors and variable rate fertilizer spreaders, and numerous companies have been formed to provide data acquisition, mapping and information management services to precision agriculture clients.

Precision agriculture is an approach for sub-dividing fields into small, relatively homogeneous management zones (Carr et al 1991, Larson & Robert 1991, Mulla 1991, 1993) where fertilizer, herbicide, seed, irrigation, drainage or tillage are custom-managed according to the unique mean characteristics of the management zone. These mean characteristics may be obtained from soil tests for nutrient availability, yield monitors for crop yield, soil samples for organic matter content, information in soil maps, or ground conductivity meters for soil moisture.

According to customary practice by agricultural consultants, fields are manually sampled along a regular grid at sample spacings ranging from 60–150 m, and the samples are analysed for desired properties. The results of these analyses are interpolated to unsampled locations by inverse distance or geostatistical techniques, and the interpolated values are classified using geographical information systems (GIS) techniques into a limited number of management zones (Mulla 1991, 1993). The boundaries of management zones are then visualized using mapping software, and management recommendations are developed for each zone. The management recommendations and mapping boundaries are stored on a computer chip which is placed in a computer on board a tractor, spreader, or seeder able to vary management according to its location in the field and the map boundaries.

The customary practices of implementing precision agriculture by agricultural consultants have several limitations. The main limitation is the adequacy of the sampling program. In high value irrigated crops such as potatoes and berries, the sampling strategy often involves more intensive sampling (60 m spacings) than in low value rain-fed crops such as wheat and corn (120 m spacings). The adequacy of the spatial sampling and interpolation strategy for 120 m grid spacings in wheat and corn cropping systems is suspect, especially given that soil survey maps in such areas typically show changes in soil mapping units and landscape position that occur every 100 m or less.

On the basis of extensive analysis of spatial structures in rain-fed agriculture, it appears that the maximum sample spacing for regular grid sampling is 60 m (Mulla 1991, 1993, Wollenhaupt et al 1994). At less intensive sample spacings, the accurate delineation of management zone boundaries becomes difficult. As this intensity of sampling is too expensive for many producers, new approaches are needed for targeted sampling. Targeted sampling may be guided by preliminary information about the site from remote sensing, soil maps, or terrain maps.

Remote sensing is currently one of the approaches attracting the most interest from scientists interested in targeted sampling. Remote sensing as discussed here involves measurement of the energy reflected or emitted from a bare soil or a growing crop. Platforms for remote sensing include satellites such as Landsat, JERS and SPOT, as well as aircraft equipped with cameras or videos. The current spatial resolution of satellite platforms is at best about 30 m (with Landsat), but future satellite platforms such as Space Imaging or Resource 21 promise resolution in the order of 1–3 m. Remote sensing has so far been used for soil mapping, terrain analysis, crop stress and yield mapping, and for estimation of soil organic matter content, among other applications. Typically, these applications have been on a scale larger than needed for precision agriculture.

Recently, there have been a few papers focusing specifically on the use of remote sensing for precision agriculture. Bhatti et al (1991) used Landsat Thematic Mapper (TM) images of bare soil to map soil organic carbon content across a large wheat field. Cokriging was used to estimate large-scale patterns in soil test phosphorus and wheat yield from the organic matter covariate. Blackmer et al (1995) used a chlorophyll sensor mounted on a tractor to measure short-wave radiation reflected from corn canopies as an indication of nitrogen stress. Anderson & Yang (1996) used multispectral videography in the colour infrared spectrum to measure within-field differences in normalized difference vegetative indices (NDVI). These differences were correlated with on-the-ground measurements of plant height, leaf area index (LAI), biomass and yield. Finally, several researchers have used tractor-mounted radiometers to detect weeds on fields prior to crop planting.

There has traditionally been little interest among practitioners of remote sensing in techniques for interpolation of data. This is because the major problem in remote sensing has typically not been the scarcity of data, but rather poor spatial resolution, large measurement errors, or difficulties in extracting information about soil properties from images taken of vegetated ground. Interest in geostatistical techniques for remote sensing applications has recently increased, due in part to the desire for quantitative models of spatial structure from remotely sensed images (Curran 1988, Woodcock et al 1988a,b). Bhatti et al (1991) used cokriging techniques along with remotely sensed soil organic carbon to estimate within-field variability of soil phosphorus fertility and crop yield. This was not only the first application of cokriging techniques to remotely sensed images, but it was also the first attempt at using remote sensing for precision agriculture. Other researchers also investigated the possibility of using cokriging techniques with remotely sensed data to reduce ground sampling intensity (Atkinson et al 1992, 1994). Finally, indicator kriging has been used by Rossi et al (1994) to estimate missing values in remotely sensed images affected by interference from cloud cover.

This paper explores the use of remote sensing for improved estimation of spatial patterns on farms managed by the techniques of precision agriculture. Two applications are presented. The first involves the use of cokriging soil phosphorus and crop yield with remotely sensed images of spatial patterns in organic matter content. The second is the use of spatial patterns in organic matter content from remote sensing to develop efficient and targeted strategies for sampling spatial patterns in soil test phosphorus.

Methods

Data for the analyses presented in this paper are for a commercial wheat farm located near St. John, Washington (Mulla et al 1992). The site has sharply rolling hills with exposed subsoils on eroded hilltops and ridges, and thick mollic epipedons at the footslope and toeslope positions. Soils were sampled in August 1987 at spacings of 15.24 m on four 655 m long transects separated by 122 m. Surface 0–30 cm samples (a composite sample of five hand auger cores) were analysed for bicarbonate-extractable phosphorus and organic matter content. Fertilizer was then applied to the field at a uniform rate of 73 kg/ha nitrogen and 6 kg/ha phosphorus. Stephens winter wheat (*Triticum aestivum*) was planted in early October 1987. In August 1988 wheat was harvested along the four

655 m long transects at intervals of 15.24 m in plots 0.6 m wide and 7–10 m long. After threshing, grain yield was determined for each location.

A Landsat TM image of bare soil at the study site was obtained for July 17 1985 (Bhatti et al 1991). TM data in 28.5 m × 28.5 m blocks were corrected to universal transverse mercator (UTM) coordinates. The ratio of TM bands 5:4 was used to estimate soil organic matter using a calibration curve obtained from several other sites in eastern Washington by Frazier & Cheng (1989).

Geostatistical analyses were conducted on soil test phosphorus, grain yield, soil organic matter content, and remotely sensed organic matter content (Bhatti et al 1991). This analysis consisted of fitting spherical models for semivariograms to soil test phosphorus, grain yield, and organic matter content. Also, spherical cross-semivariogram models were developed to describe the spatial dependence between soil test phosphorus/soil organic matter or grain yield/soil organic matter. Block kriging and cokriging techniques (Burgess & Webster 1980, McBratney & Webster 1983, Isaaks & Srivastava 1989) were used to estimate values for soil test phosphorus, grain yield and organic matter content on 15.225 m × 30.48 m grids for ease of comparison with remotely sensed organic matter estimates.

For evaluation of typical agricultural consulting practice, the soil test phosphorus dataset consisting of 172 sampling locations was pared down to 24 samples collected on a 120 m regular grid spacing. This dataset was kriged on a 15.225 m × 30.48 m grid using the semivariogram for the full dataset. For evaluation of the potential for targeted sampling, a second dataset was produced from the original soil test phosphorus data. Data were extracted from 51 of the 172 data points at locations on the remotely sensed organic matter map which corresponded to subregions having low (<1.5%), moderate (1.5–2.5%), or high (>2.5%) organic matter contents. This subset of 51 targeted sampling locations was kriged to produce soil test phosphorus values on a 15.225 m × 30.48 m grid.

Each of the kriged or cokriged maps were classified using a customized GIS (Mulla 1991, 1993) into three management zones. When classifying maps of organic matter content, the cut-offs 1.5% and 2.5% were used to separate the data into three distinct management zones. When classifying maps of soil test phosphorus, we used 5 and 10 ppm cut-off points to separate the data into three distinct management zones. The cut-offs for soil test phosphorus are based upon fertilizer recommendations for eastern Washington state (Halvorson et al 1982). According to these recommendations, the zone with <5 ppm phosphorus should receive 53.5 kg/ha P_2O_5, the zone with between 5 and 10 ppm should receive 35.7 kg/ha P_2O_5, and the zone with greater than 10 ppm should receive no phosphorus fertilizer.

Results

Classical statistics

Ground-sampled organic matter contents averaged 2%, about 1% lower than the mean organic matter content estimated by remote sensing. This difference is probably attributable to the lack of calibration at this study site for the model used to estimate organic matter from the remotely sensed digital numbers for band ratio 5:4. The coefficient of variation (CV) for soil organic matter content (41.3%) was significantly larger than the CV for remote sensing estimates of organic matter (28.2%). This is most likely the result of the spatial averaging inherent in the remotely sensed estimates, which tends to decrease variability.

Field-averaged soil test phosphorus was about 15 ppm, which according to fertilizer guides for winter wheat (Halvorson et al 1982) indicates no phosphorus deficiencies if the field were to be uniformly fertilized at a single rate. The CV for soil phosphorus is 50%, which indicates significant variability about the mean value, and hints at the possibility of portions of the field that are deficient in phosphorus. Grain yield in the field averaged roughly 4000 kg/ha, with a moderate CV of about 29%.

Spatial structure

Spherical semivariogram and cross-semivariogram models provided an excellent description of the spatial dependence for measured properties (Table 1). The range or sill of the spherical models for soil organic matter content and remotely sensed organic matter content were nearly identical, even though very different measurement techniques were used to estimate spatial patterns. Nugget values for organic matter content semivariograms were smaller with remote sensing estimates than ground-based measurements, suggesting that measurement errors are less with remote sensing than with soil sampling. Cross-semivariograms between organic matter and either grain yield or soil phosphorus (Table 1) indicated well developed co-spatial dependence as indicated by the values for the range equivalent to up to 10 sample spacings, and the large proportion of variance between the sill and nugget values.

Interpolated spatial patterns

Spatial patterns for kriged organic matter based upon 172 soil sampling locations (Fig. 1) divide the field into three large and distinctly different sub-sections. In contrast, the

TABLE 1 Summary of parameters for spherical semivariogram and cross-semivariogram models at the St. John site

Property	*Nugget*	*Sill*	*Range*
Organic matter (%)	0.14	0.72	114
Landsat organic matter (%)	0.03	0.68	114
Soil phosphorus (ppm)	27.58	63.38	145
Grain yield (kg/ha)	840000	1140000	70
Organic matter (%)–grain yield (kg/ha)	160.0	540	120
Organic matter (%)–soil phosphorus (ppm)	0.4	4.1	107

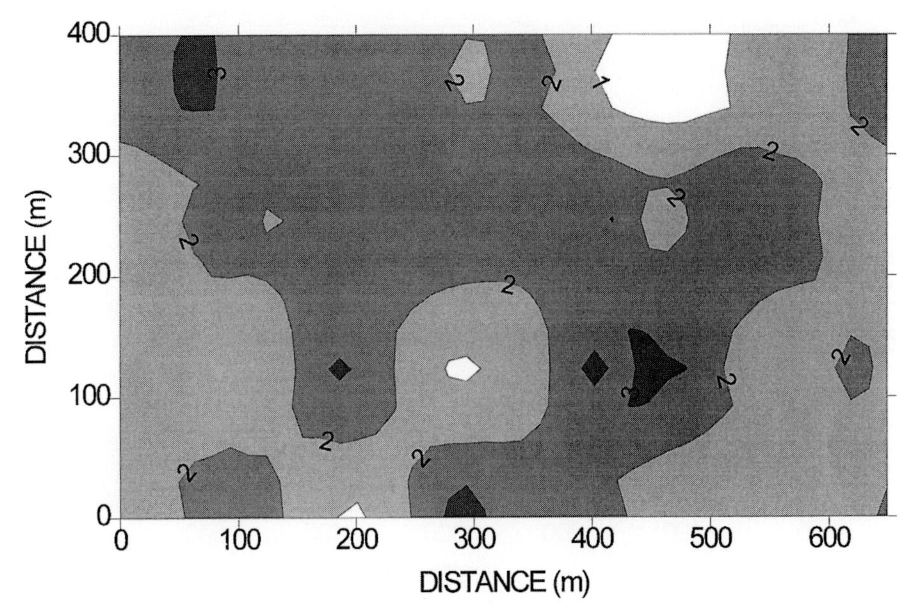

FIG. 1. Kriged organic matter (%).

spatial patterns for remotely sensed organic matter content (Fig. 2) differentiate the field into about seven distinctly different sub-sections. Thus, satellite imagery of bare soil provides more detail about spatial patterns in organic matter content than intensive ground sampling. By classifying these satellite estimates of spatial patterns based on management zone cut-off levels for low (<1.5%), moderate (1.5–2.5%), (>2.5%) and high organic matter contents, five patches with moderate to low levels of organic matter can be delineated (Fig. 3). These patches are the locations where we hypothesize that phosphorus deficiencies will occur, and where moderate to high rates of phosphorus fertilizer will be required to attain optimum soil fertility levels.

Spatial patterns in kriged wheat yields (Fig. 4) show yields ranging from about 1000–6000 kg/ha in a 26 ha region. Low yielding locations correspond relatively closely with regions having lower organic matter contents (Fig. 2). Portions of the field with lower organic matter contents are locations with eroded subsoils at upper slope positions. Wheat yields estimated by cokriging with remotely sensed organic matter content (Fig. 5) show details that are not evident in the kriged spatial patterns. In particular, a large contiguous zone of high yields is apparent centred on the coordinates (350, 100). Thus, cokriging with remotely sensed organic matter seems to be a powerful method of extracting spatial information in crop yields from information provided by ground-based measurements.

Spatial patterns for kriged soil test phosphorus (Fig. 6) show values less than 5 ppm that are located roughly in accordance with organic matter contents that are less than 2% (Fig. 2). Soil test phosphorus levels higher than about 10 ppm do not always appear

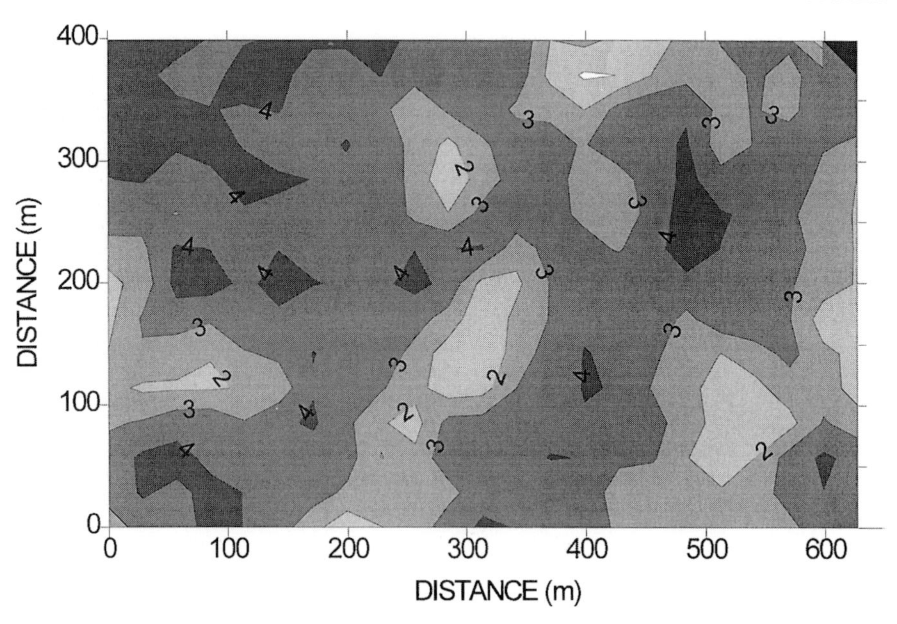

FIG. 2. Satellite-estimated organic matter (%).

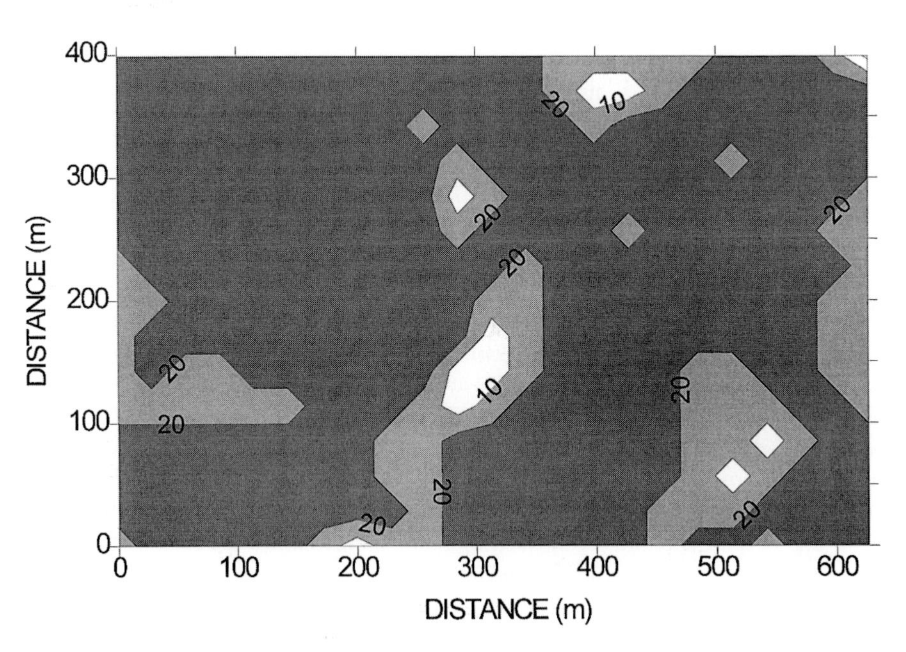

FIG. 3. Organic matter-based classification (0–10, <1.5%; 10–20, 1.5–2.5%; 20–30, >2.5%).

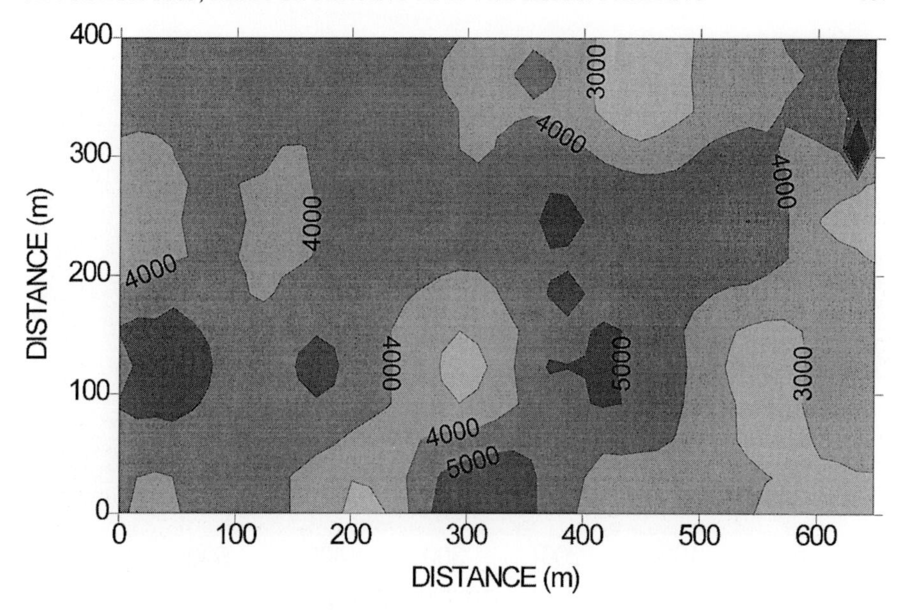

FIG. 4. Kriged wheat yield (kg/ha).

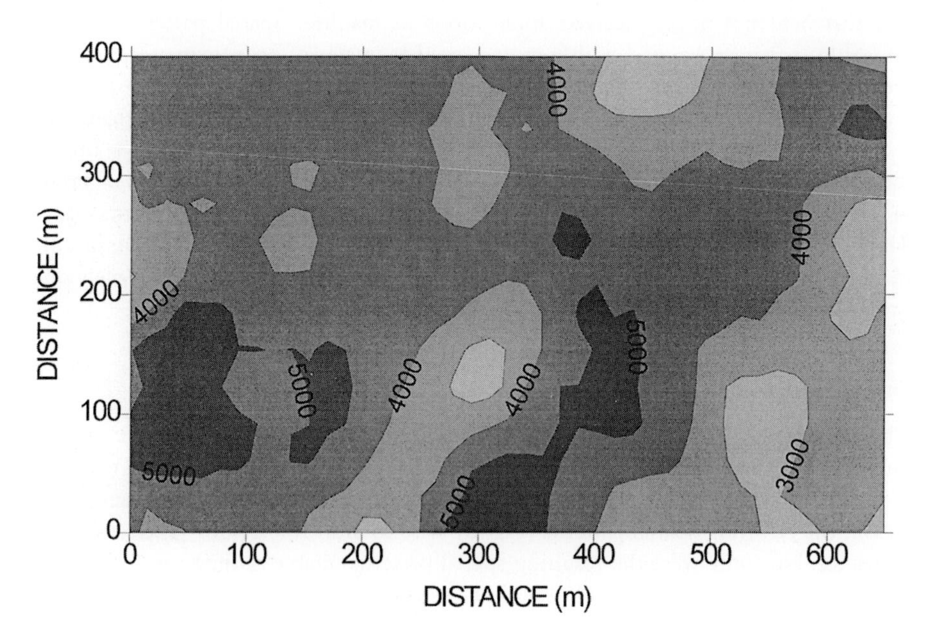

FIG. 5. Cokriged wheat yield (kg/ha).

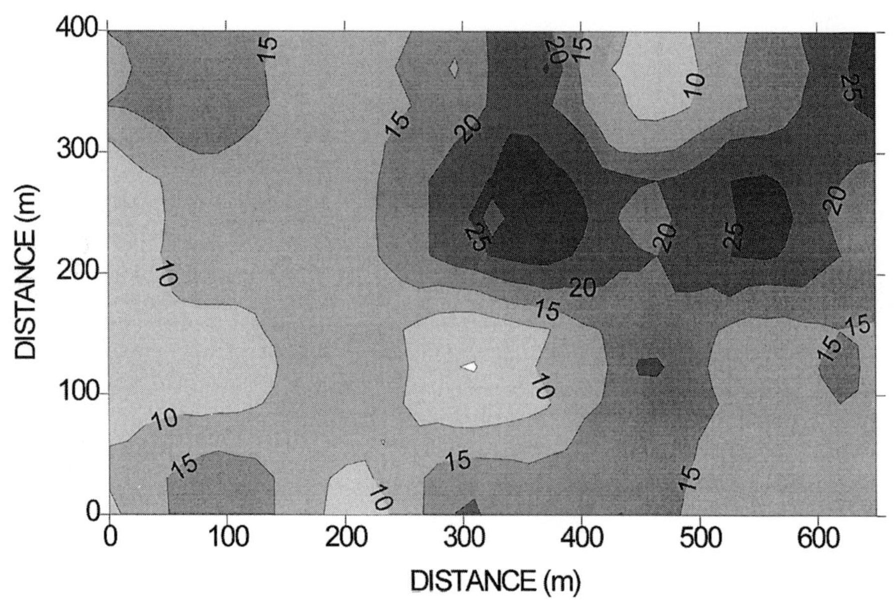

FIG. 6. Kriged soil test phosphorus levels (ppm).

to be spatially consistent with patterns in organic matter content. It should be noted that this field has never received applications of manure. Spatial patterns in soil phosphorus from cokriging with remote-sensed estimates of organic matter content are shown in Fig. 7. Many of the locations in the cokriged map with soil phosphorus levels below 5 ppm are consistent with those in the kriged map, however, there are locations with low phosphorus levels in the cokriged map that did not show up in the kriged map. Similarly, there are locations with phosphorus levels above 25 ppm in the cokriged map that were not correctly identified in the kriged phosphorus map. Thus, the use of cokriging from satellite images shows details for spatial patterns in soil phosphorus that are not possible to delineate in the kriged map that is based solely on ground sampling.

Evaluation of grid and targeted sampling strategies

The typical practice of precision farming consultants in rain-fed regions is to sample fields on a regular grid at spacings of approximately 120 m. After kriging 24 of the original 172 soil test phosphorus values selected from a regular grid at spacings of 120 m, the resulting spatial patterns (Fig. 8) were found to be a poor representation of reality. Not only were the resulting spatial patterns unable properly to locate the regions of high soil test phosphorus, but there were no regions on the map with phosphorus levels lower than the 10 ppm at which phosphorus deficiencies are indicated. Thus, sampling on a 120 m grid for this field would have resulted in a

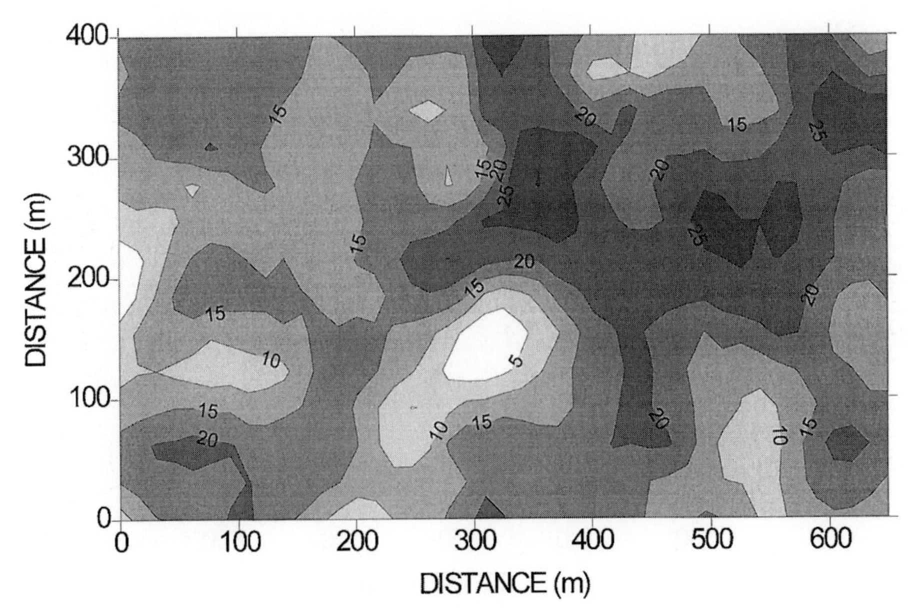

FIG. 7. Cokriged soil test phosphorus levels (ppm).

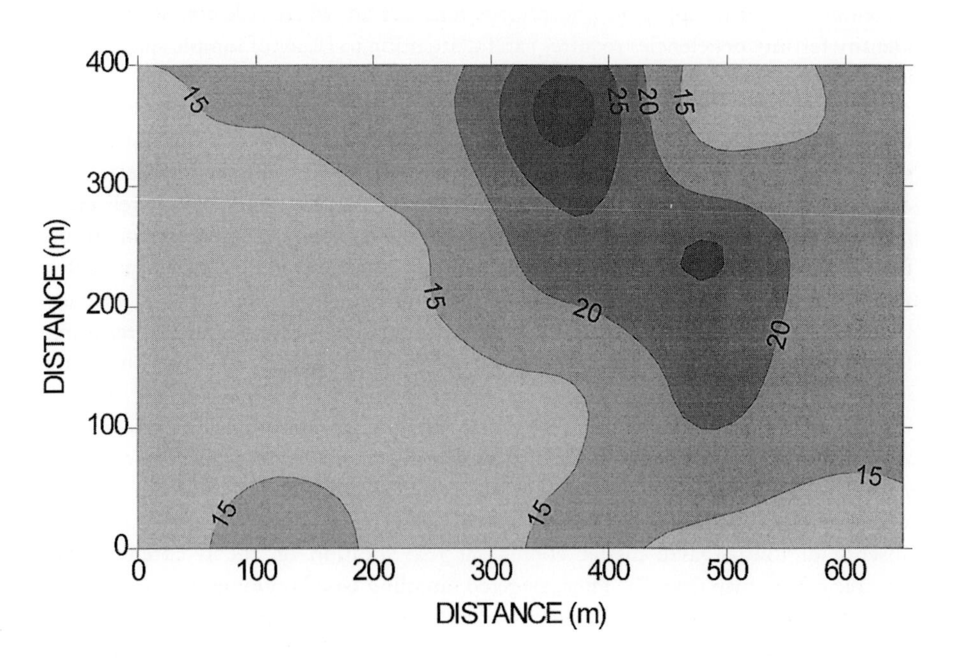

FIG. 8. Grid-sampled kriged phosphorus (ppm).

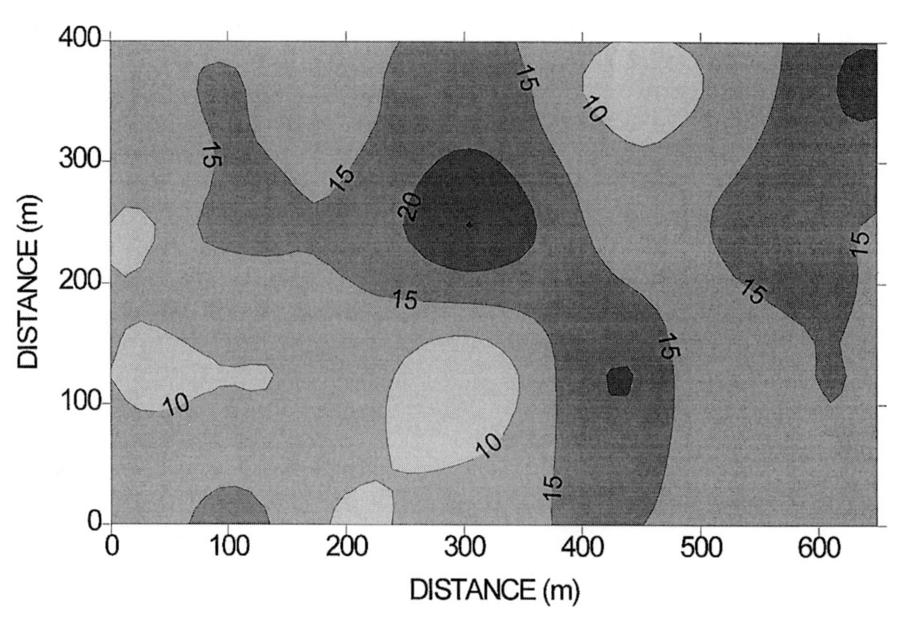

FIG. 9. Targeted sampling kriged phosphorus (ppm).

recommendation to apply no phosphorus fertilizer anywhere. Clearly, sampling to identify fertility deficiencies requires careful attention to issues of sample spacing and location.

To sample the field efficiently we propose a targeted sampling scheme which is based upon spatial patterns in remotely sensed organic matter content (Fig. 2). In this strategy, sampling locations are selected where transitions from moderate to low organic matter contents (or vice-versa) occur. In regions with relatively homogeneous and moderate to high organic matter contents, a sample spacing of up to 100 m is selected. Thus, in the targeted sampling scheme, sample spacings are not regular, and regions with low organic matter contents are sampled more intensively than any other region. The spatial patterns in kriged soil test phosphorus resulting from this targeted sampling scheme are shown in Fig. 9. These spatial patterns are not as complicated as those obtained by cokriging with remotely sensed organic matter (Fig. 7). However, the locations having low soil test phosphorus levels are remarkably consistent between the kriged targeted sampling map and the cokriged map. The one exception is the failure to identify the low soil test phosphorus levels with targeted sampling at the coordinates surrounding (550, 50). This is due to the absence of low organic matter content at that location on the remotely sensed organic matter map (Fig. 2). Thus, targeted sampling based upon remote sensing has the potential for providing reasonably accurate nutrient deficiency maps in eastern Washington with reduced sampling effort as compared with sampling schemes based only on intensive soil sampling. This approach assumes spatial correlation between soil

TABLE 2 Areal Coverage (ha) by fertilizer management zones as a function of sampling and interpolation methods

Sampling and interpolation method	*Management zone C (high fertilizer)*	*Management zone B (moderate fertilizer)*	*Management zone A (no fertilizer)*
Intensive, cokriging	0.6	2.4	23.0
Intensive, kriging	0.1	3.7	22.2
Targeted, kriging	0	2.5	23.5
Grid, kriging	0	0	26.0

organic matter and soil phosphorus, which may not occur in areas where manure is soil-applied.

What is really critical in precision agriculture is having a map of spatial patterns that accurately identifies the locations and areal extent of distinct management zones within the field. The kriged and cokriged spatial patterns for soil test phosphorus were reclassified according to published cut-off levels used in making phosphorus fertilizer recommendations (Halvorson et al 1982). Zones with soil test phosphorus levels less than 5 ppm should receive relatively high rates of fertilizer, zones with levels between 5 and 10 ppm should receive moderate rates, and zones with greater than 10 ppm should receive no fertilizer.

Table 2 shows the area in ha for each of the three management zones described above. Zone C, which receives the highest rates of fertilizer, has an area of 0.6 ha by cokriging, an area of 0.1 ha by kriging, an area of 0 ha by targeted sampling and an area of 0 ha by grid sampling. Zone B, which receives moderate rates of fertilizer, has an area of 2.4 ha by cokriging, an area of 3.7 ha by kriging, an area of 2.5 ha by targeted sampling and an area of 0 ha by grid sampling. Zone A, which receives no fertilizer, has an area of 23 ha by cokriging, an area of 22.2 ha by kriging, an area of 23.5 ha by targeted sampling and an area of 26 ha by grid sampling. For comparison the zones with low, moderate and high organic matter content (Fig. 3) have areas of 0.8, 5.3 and 19.9 ha, respectively. Based on this information, the use of intensive ground sampling with interpolation by cokriging remotely sensed organic matter estimates provides the best overall accuracy in delineating management zones. The use of intensive ground sampling with kriging and the use of targeted sampling with kriging both result in the underestimation of area requiring high rates of phosphorus fertilizer, but give satisfactory estimates of the area requiring moderate rates of fertilizer. The grid sampling method is totally inadequate at this study site for delineating fertilizer management zones.

Besides the importance of properly estimating the areal coverage for each management zone, it is important that the zones be properly located on the landscape. The locations for each management zone were plotted as contour maps using contour intervals of 0–10 for management zone C (severe P deficiencies),

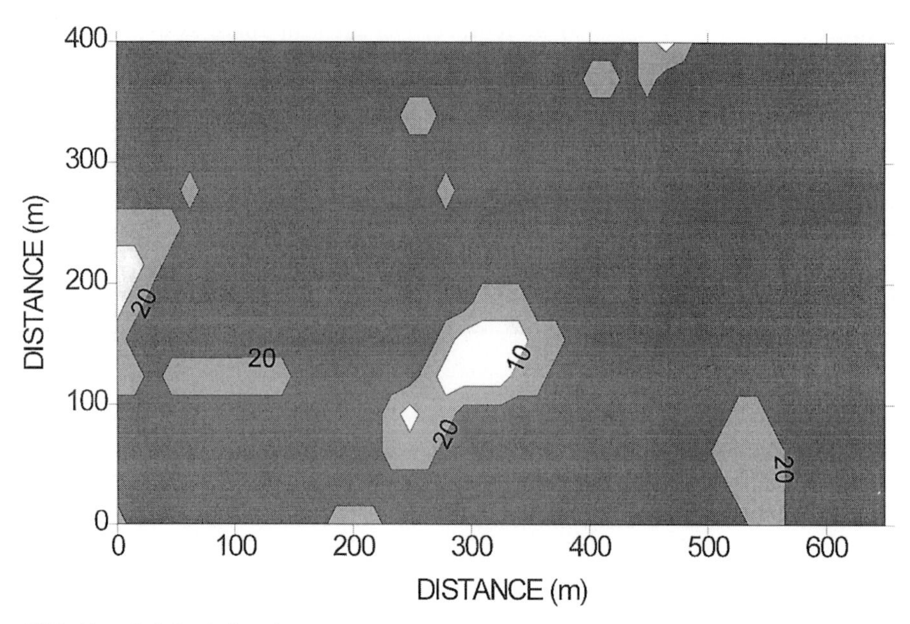

FIG. 10. Cokriged phosphorus management zones.

intervals of 10–20 for zone B (moderate P deficiencies), and intervals of 20–30 for zone A (no P deficiencies). Fig. 10 shows the location of the three phosphorus management zones based upon intensive soil sampling (172 sample locations) and cokriging with remotely sensed organic matter contents. This map delineates four distinct regions which are severely deficient in P, 10 regions which are moderately deficient, and the remainder of the field is not deficient in P. Fig. 11 shows the P management zones from intensive soil sampling and kriging. There is now only one small area delineated with severe P deficiencies and four large areas with moderate deficiencies. The locations of the moderately deficient zones correspond well with the locations delineated in the cokriged map (Fig. 10). Fig. 12 shows P management zones from a targeted sampling strategy (51 sample locations) based upon variations in remotely sensed organic matter content (Fig. 2). The targeted sampling strategy shows five regions with moderate P deficiencies, whose locations correspond well with those in the management zone maps based upon cokriging (Fig. 10) and kriging (Fig. 11). The targeted sampling strategy tends to underestimate the area requiring moderate rates of fertilizer in the region near coordinates (50, 150), but otherwise is satisfactory for delineating phosphorus management zones. As these examples show, there is a trade-off between accuracy in mapping management zones and the intensity of sampling. Best accuracy is achieved by using the more intensive soil sampling strategy, but satisfactory accuracy can be achieved by using reduced numbers of soil samples which are spatially targeted to critical locations using a remotely sensed image of bare soil.

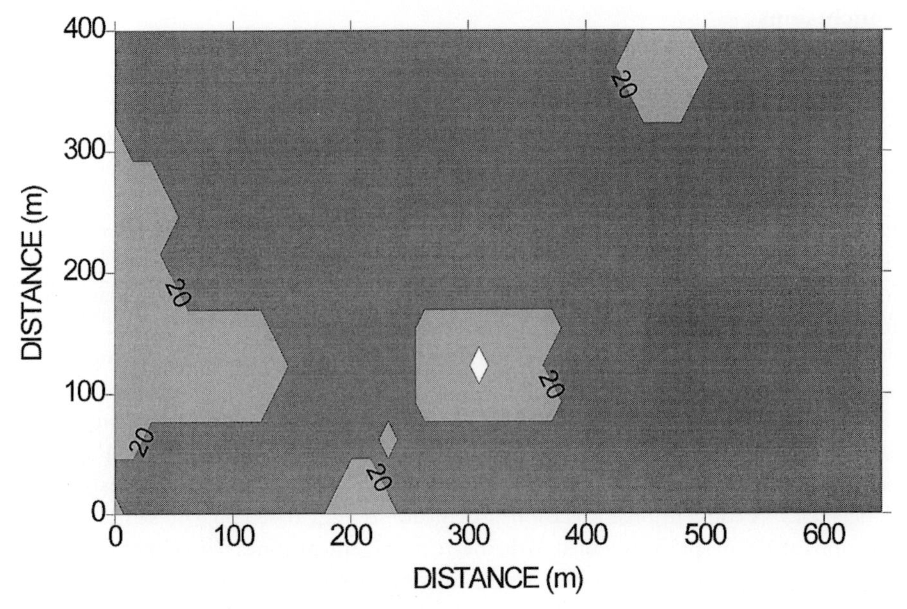

FIG. 11. Kriged phosphorus management zones.

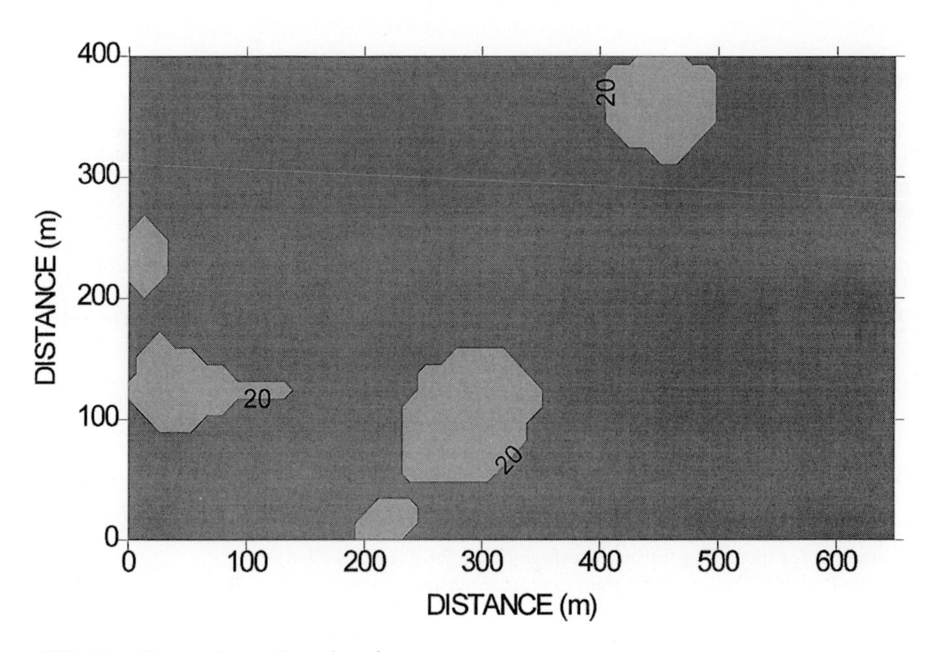

FIG. 12. Targeted sampling phosphorus management zones.

Conclusions

In this paper we have demonstrated that spatial patterns in soil test phosphorus or grain yield are spatially correlated with spatial patterns in soil organic matter content. At this study site, this spatial co-dependence resulted from patterns of erosion on steep hills and knobs which exposed subsoils and decreased topsoil depth. Spatial patterns in soil organic matter content could be assessed with reasonable accuracy using Landsat band ratios 5:4 at a spatial resolution of 28.5 m × 28.5 m. Estimation of spatial patterns in soil test phosphorus was more accurate with the use of intensive soil sampling (172 locations) and cokriging with remotely sensed organic matter contents, than with intensive soil sampling and kriging techniques. Remotely sensed organic matter contents were used to develop a targeted sampling strategy in which 51 locations with low to moderate organic matter contents were sampled more intensively than locations with moderate to high contents. Spatial patterns delineated with kriging of targeted samples were nearly as accurate as those based upon intensive sampling and cokriging, but were unable properly to identify small regions with severe phosphorus deficiencies. A conventional precision agriculture strategy based on grid sampling (24 locations at 120 m spacings) failed to delineate management zones with either severe or moderate phosphorus deficiencies. Thus, the combined use of remotely sensed images of bare soil, targeted soil sampling and kriging techniques appears to have great potential as a reasonably cost-effective methodology for assessing spatial patterns of phosphorus deficiencies in wheat fields of eastern Washington state.

References

Anderson GL, Yang C 1996 Multispectral videography for site-specific farm management. In: Huberty B, Lurie JB, Caylor JA, Coppin P, Robert PC (eds) Multispectral imaging for terrestrial applications. Proc Soc Photo Optical Instr Eng, SPIE, Bellingham, WA, p 79–86

Atkinson PM, Webster R, Curran P J 1992 Cokriging with ground-based radiometry. Remote Sens Environ 41:45–60

Atkinson PM, Webster R, Curran P J 1994 Cokriging with airborne MSS imagery. Remote Sens Environ 50:335–345

Bhatti AU, Mulla DJ, Frazier BE 1991 Estimation of soil properties and wheat yields on complex eroded hills using geostatistics and Thematic Mapper images. Remote Sens Environ 37:181–191

Blackmer TM, Schepers JS, Meyer GE 1995 Remote sensing to detect nitrogen deficiency in corn. In: Robert PC, Rust RH, Larson WE (eds) Site-specific management for agricultural systems. American Society of Agronomy, Madison, WI, p 505–512

Burgess TM, Webster R 1980 Optimal interpolation and isarithmic mapping of soil properties. I. The semivariogram and punctual kriging. J Soil Sci 31:315–331

Carr PM, Carlson GR, Jacobsen JS, Nielsen GA, Skogley EO 1991 Farming soils not fields: a strategy for increasing fertilizer profitability. J Prod Agric 4:57–61

Curran PJ 1988 The semivariogram in remote sensing: an introduction. Remote Sens Environ 24:493–507

Frazier BE, Cheng Y 1989 Remote sensing of soils in the eastern Palouse region with Landsat thematic mapper. Remote Sens Environ 28:317–325

Halvorson AR, Koehler FE, Engle CF, Morrison KJ 1982 Fertilizer guide: dryland, wheat general recommendations. Coop Ext Svc FG0019, Washington State University Press, Pullman, WA

Isaaks EH, Srivastava RM 1989 Applied geostatistics. Oxford University Press, New York

Larson WE, Robert PC 1991 Farming by soil. In: Lal R, Pierce FJ (eds) Soil management for sustainability. Soil Water Conservation Society, Ankeny, IA, p 103–112

McBratney AB, Webster R 1983 Optimal interpolation and isarithmic mapping of soil properties. V. Coregionalization and multiple sampling strategy. J Soil Sci 34:137–162

Mulla DJ 1991 Using geostatistics and GIS to manage spatial patterns in soil fertility. In: Kranzler G (ed) Automated agriculture for the 21st century. American Society of Agricultural Engineers, St Joseph, MI, p 336–345

Mulla DJ 1993 Mapping and managing spatial patterns in soil fertility and crop yield. In: Robert PC, Rust RH, Larson WE (eds) Soil-specific crop management. American Society of Agronomy, Madison, WI, p 15–26

Mulla DJ, Bhatti AU, Hammond MW, Benson JA 1992 A comparison of winter wheat yield and quality under uniform versus spatially variable fertilizer management. Agric Ecosys Environ 38:301–311

Rossi RE, Dungan JL, Beck LR 1994 Kriging in the shadows: geostatistical interpolation for remote sensing. Remote Sens Environ 49:32–40

Wollenhaupt NC, Wolkowski RP, Clayton MK 1994 Mapping soil test phosphorus and potassium for variable-rate fertilizer application. J Prod Agric 7:441–448

Woodcock CE, Strahler AH, Jupp DLB 1988a The use of variograms in remote sensing. I. Scene models and simulated images. Remote Sens Environ 25:323–348

Woodcock CE, Strahler AH, Jupp DLB 1988b The use of variograms in remote sensing. II. Real digital images. Remote Sens Environ 25:349–379

DISCUSSION

McBratney: A couple of points of clarification. First, what was the relationship between organic matter and phosphorus in this field? Second, what was the basis of the targeted sampling?

Mulla: The relationship between phosphorus and organic matter was reasonably good from classical statistical regression: the coefficient of linear regression (r value) was about 0.56.

In the targeted sampling we tried to select a subset of our 172 soil samples where organic matter content was at or below 1.5% and where a transition to moderate values (between 1.5 and 2%) was occurring. We targeted these regions more intensively than the regions with higher organic matter content. The targeted samples were located in the regions that had organic matter content typically lower than about 1.5%.

Voltz: When you use targeted observations for interpolation, isn't there a problem of bias? Your kriged maps will systematically underestimate organic matter content in the regions with high content because you selected observations only from regions which you assumed to have low organic matter content. To avoid this bias, a better solution would perhaps be not to interpolate across these regions with assumed high organic matter content.

Mulla: There is some bias in the targeted sampling, but in this field there was a very good relationship between organic matter content and phosphorus. In many fields where manure is applied this is not the case, but this field did not have any applications of manure. I think this kind of approach would probably fail where manure is applied because the variations in organic matter content may not be the only factor that really drives the variability in phosphorus. Also, to some extent each region has a different set of characteristics that need to be evaluated, and what I'm proposing is that for each region people should look for features that are easily visible either in soil maps, terrain attributes or aerial photos and try to develop a good understanding of the factors that control the variability in each property.

Voltz: Wouldn't it be as simple just to say from the beginning that these are the management sites you are focusing on and then just to map the phosphorus content over the management sites that you suspect to be deficient in phosphorus?

Mulla: Yes. In an earlier paper (Mulla 1993) I actually advocated that approach, which is to delineate areas that are low in organic matter content, and go in and just sample those areas to determine what the phosphorus content is, and then to develop your management systems based on a limited sampling. This is a slightly different way of doing the targeted sampling.

Voltz: Did you do some validation of your maps with independent data? Sometimes you argue that you see more details on certain maps than on others, but the question is whether these details are real or not. Of course the answer to this question depends on the confidence intervals you have on your predictions.

Mulla: I didn't actually do any validation on this data set.

Bouma: We deal quite a bit with what we have called 'research negotiation'. If you offer farmers advice you cannot take it for granted that they will take it: they will want to see alternatives. In your paper, the company grid (120 m grid spacings) showed the areas that were deficient in phosphorus. If the farmer follows that advice, is the additional value he gets out of that system worth more than the cost of detailed sampling? If that is the case it's a good selling point for your procedure. How do these costs compare?

Mulla: We did a cost analysis: as I remember, there was a benefit at this site to precision application of fertilizer of about US$27 per acre. This benefit was estimated from the cost savings involved in applying less fertilizer for precision management versus conventional uniform management.

Bouma: What would the cost analysis be with the 120 m grid spacing map, which didn't show the deficient spots?

Mulla: They wouldn't have applied fertilizer so they would have received a penalty in terms of yield.

Groffman: Is that really true, though? Weren't the areas that were deficient in phosphorus the ridge tops, which had shallow soil and were dry, low yielding areas anyway?

Mulla: Sure, but there is still a significant response in yield at those locations. From the viewpoint of profitability, there is a good reason for doing accurate soil sampling. The returns often are justified on the basis of the cost of sampling.

Webster: On this question of cost, in Britain at least the break-even point is around one sampling point per hectare. If you sample more than that, then in general the cost of the sampling is greater than the benefit you can expect from more precise application of fertilizer.

With too few data, however, you encounter a problem of using kriging or any other method of interpolation. If you have only 29 points, or even 51, the variograms you get from them are so crude that you are unlikely to be able to get decent estimates. In your research exercise, Dr Mulla, you had 172 points, and I take it you used those variograms on your 29 data. But if you had used only 29 data, then you would have been in big trouble.

Mulla: I agree; I did use the variogram from the full dataset.

Robert: When the concept of precision agriculture was developed in the mid 1980s, it was called 'farming by soil type'. The intention was to use mapping units of the USDA-NRCS (US Department of Agriculture-National Resource Conservation Service) county soil survey (1:20 000 scale) as management units and soil sampling units. However, it appeared quickly that because of the impact of past nutrient management practices, a more intensive soil sampling procedure was needed. At that time a systematic grid sampling was the only manageable approach for agriconsultants, and pseudoeconomics guided the grid size with little understanding of the effect on the quality of the results. Now, progressively, with a better understanding of spatial variability from yield maps and the availability of differential global positioning system (DGPS), the value of a directed or smart soil sampling is progressively recognized and should become more frequently adopted. Soil sampling on a 2.5 acre (1 ha) grid down to 4 feet (1.2 m) is a common practice in the sugar beet growing region of NW Minnesota. In this region, about 80% of producers are using precision nutrient management. This can be justified by a net return of between $40–70 per acre ($100–170 per ha) above conventional management due to a greater sugar content resulting from a better N management. However, some farmers have recently found that the microrelief present in that region—less than a 2 foot (0.6 m) difference in elevation—has an important influence on the N mineralization, and have started precise topographic surveys of fields for directed soil sampling and better application of N.

Groffman: Where are these big returns coming from? Do they come from increased yield or decreased costs?

Robert: For sugar beet, the quality of the crop—that is, the sugar content—is the most important economic criterion. It is the basis of the payment to the farmers. As I mentioned earlier, the sugar content is related principally to the N management. A more precise N management increases the sugar content. If there is too much N available after approximately August 1st, foliage is produced instead of sugar storage. For other crops such as corn and soybean, increased returns may come from both increased yields and decreased input costs, including lower drying costs. In the future, crop quality may also become an important factor for grain crops.

Rabbinge: In many arable crops the fine-tuning of nitrogen is vitally important. In potatoes there is a penalty because too much nitrogen produces more vegetative material and fewer, smaller potatoes. In wheat, excess nitrogen causes an upsurge in disease epidemics and very often lodging. This is why fine-tuning in time in relation to fine-tuning in space in such crops is so vitally important.

Goense: In your experiment, the farmer seemed to accept the number of 24 samples. Would it help to use cokriging with 24 samples and would it make sense to target that sampling just with 24 samples?

Mulla: We used cokriging with 51 target samples and it worked adequately. I'm a little hesitant to drop the number of target sampling sites much below that, although I don't know whether it's feasible or not. My intuition is that if you reduce the number of target samples to 24 you might lose even more accuracy. I have a feeling that we're going to probably do an inadequate job of characterizing the spatial patterns with 24 samples at this site.

Barnett: I wanted to ask you about the targeting. Your targeting was 51 out of a specific subset of 172. It was not an optimal targeting, that is, a choice of the 51 best sites for maximal information. Were you able to do any sort of sensitivity analysis to determine what you might have been losing by targeting within a prescribed set of 172, rather than arbitrarily targeting 51 over the field?

Mulla: There's no doubt that we could have done a better job of targeting if we had not restricted ourselves to the 172 samples. Obviously, that's the approach that should be taken in further applications of this procedure.

Barnett: Do you have any feel for what sort of loss there was from not doing that?

Mulla: By comparing the targeted sampling approach (51 samples) with the intensively sampled soil phosphorus data (12 samples) I think we do have a sense of that. We showed that it was possible accurately to identify the phosphorus-deficient zones within the field using a targeted sampling approach based on remote sampling and cokriging.

Stafford: We have taken an approach to target sampling based on yield maps. We take a sequence of yield maps and apply a clustering analysis to identify areas that perform similarly from year to year (Lark & Stafford 1997). We've applied this technique to about a dozen fields and we have got sequences of yield maps and have identified between two and six clusters in each field. We can then target into those clusters for soil sampling. This is another approach, and it has the benefit that yield mapping is already available to many farmers and so it's not an expensive data collecting exercise.

Mulla: That is a viable approach. However, when you divide a field on the basis of yield response, you are not always able to make the optimum delineation of the field with regard to, for instance, zones that are deficient in phosphorus. This is the only drawback that you have to be aware of: the yield classes do not necessarily correspond in the optimal sense with the classes that you would obtain for, say, soil test phosphorus based upon soil sampling.

Stafford: One could investigate clusters and seek to identify the limiting factors.

References

Lark RM, Stafford JV 1997 Some methods for analysing temporal variation in yield maps. Ann Appl Biol, in press

Mulla DJ 1993 Mapping and managing spatial patterns in soil fertility and yield. In: Robert PC, Rust RH, Larson WE (eds) Soil specific crop management. Agronomy Soc Am, Madison, WI, p 15–26

Space–time statistics for decision support to smart farming

A. Stein, M. R. Hoosbeek and G. Sterk*

*Department of Soil Science and Geology, PO Box 37, Wageningen Agricultural University, 6700 AA Wageningen and *Department of Irrigation and Soil and Water Conservation, Agricultural University, Nieuwe Kanaal 11, 6709 PA, Wageningen, The Netherlands*

Abstract. This paper summarizes statistical procedures which are useful for precision farming at different scales. Three topics are addressed: spatial comparison of scenarios for land use, analysis of data in the space–time domain, and sampling in space and time. The first study compares six scenarios for nitrate leaching to ground water. Disjunctive cokriging reduces the computing time by 80% without loss of accuracy. The second study analyses wind erosion during four storms in a field in Niger measured with 21 devices. We investigated the use of temporal replicates to overcome the lack of spatial data. The third study analyses the effects of sampling in space and time for soil nutrient data in a Southwest African field. We concluded that statistical procedures are indispensable for decision support to smart farming.

1997 Precision agriculture: spatial and temporal variability of environmental quality. Wiley, Chichester (Ciba Foundation Symposium 210) p 120–133

A modern approach towards smart farming requires careful collection and use of data. These data are used to characterize the land as a four-dimensional unit (three spatial dimensions and time) to model crop growth, as well as to predict the effects of fertilizer application. Smart farming relies on the extensive use of data to feed agricultural models. Commonly data have to be collected from field observations. Precise modelling requires intensive sampling, whereas budgets and common sense usually favour smaller sample sizes (Burrough 1996). The aim of smart farming is to find a balance between optimal yield, maximum profit and minimal environmental pollution. Making the most of the data requires the use of statistical procedures (Mulla 1993). In this paper we address three examples of the use of space–time statistics for smart farming.

Spatial statistics focused originally on interpolation procedures (Journel & Huijbregts 1978) to move from point data towards expressions for areas of land. As a next stage, patterns on interpolated maps need to be compared (Stein et al 1997, Van Uffelen et al 1997). Secondly, farmers, scientists and politicians have realized that data tend to fluctuate in space *and* time. Data in space and time, however, are usually anisotropic, with a cause–effect relation in time. Modern developments in spatial statistics now permit the extension of the random field approach, that was so successful in spatial statistics, to the space–time domain (Christakos 1992). An important issue is sampling in space and in time. In particular, the classical experimental design was never designed for interpolation procedures. An alternative is presented and its properties will be evaluated using simulated random fields.

Comparing land use scenarios

Smart farming aims to maximize production, while reducing costs due to fewer applications and minimizing environmental problems. These are possibly conflicting goals, because a uniform fertilizer application level that will maximize production on the location with the worst nutrient status, may cause leaching on locations with a better nutrient status. In smart farming it is crucial to evaluate different options for land use, particularly if decisions have to be made concerning the amounts of nitrate, fertilizer and pesticide to apply. A geographical information system combined with appropriate models and geostatistical procedures provides a reliable basis for decision making.

Example

A study was carried out in the 1990s on a field in the Wieringermeer area, The Netherlands. Six different scenarios for nitrate application (Table 1) were compared with each other and also with the current fertilizing practice in the Netherlands (Finke & Stein 1994). Scenarios are usually compared by evaluating an estimate of the spatial mean to a threshold value. Here, the field average leaching concentration would have to be compared to the current threshold value of 50 mg nitrate/dm^3. This is

TABLE 1 Description of fertilizer scenarios. N_{adv} and N_{opt} are current advice and optimal levels in kg N ha^{-1}, related as $N_{adv} = N_{opt} - a \cdot N_{min}$ where a is a crop-specific factor and N_{min} is the amount of inorganic nitrate present in the rootable layer

Scenario	*Modficiation with respect to current situation*	*Number of simulations*
S_0	None	402
S_1	$N_{adv} \cdot 0.25$	Optimized
S_2	$N_{adv} \cdot 0.50$	Optimized
S_3	None	Optimized
S_4	$N_{adv} \cdot 1.5$	Optimized
S_5	$N_{adv} \cdot 2.0$	Optimized
S_6	$N_{opt} \cdot 2.0$	Optimized

not realistic, because uncertainties caused by model inaccuracies and spatial variation of the property obscure the sensible use of absolute values. Hence, we applied a probabilistic criterion: 'nowhere in the field may the probability of exceeding a leaching concentration of 25 mg nitrate/dm^3 be greater than 0.05'. With this criterion spatial statistics must be used for decision making. The purpose of this study, therefore, was to investigate the role of disjunctive kriging and cokriging (Matheron 1976, Webster & Oliver 1989).

Simulations start with two years of fertilizing (scenario S_0) with initial nitrogen amounts corresponding to the level measured in the field. Thereafter, scenarios S_1 to S_6 are applied over a period of 17 months in which three crops are grown: spring barley, a catch-crop (rye grass) and potatoes. Attention focused on the simulated nitrate concentration in the leaching water at 0.8 m. We compared the fertilizing scenarios by investigating nitrate leaching and crop production in 402 soil profiles where model input variables were determined. The model LEACHN (Wagenet & Hutson 1989) extended with a crop growth submodel was used. Simulations on all locations would result in a very high computing effort (11.2 central process usage days per scenario). We therefore reduced the number of simulations by using the simulation results from S_0 as a co-variable in disjunctive cokriging. A test set of 50 locations was selected at random to test the differences.

For scenarios S_0 to S_6, a minimal data set of 38 simulations was defined, located on a triangular grid with a base distance of 48 m to cover the entire field. If both the Mean Variance of the Prediction Error (MVPE) and Mean Squared Error for Predictions (MSEP) criteria failed to be satisfied, the data set was expanded by performing 10 more simulations, randomly chosen from the remainder of available locations. Variograms were fitted again and cross-validated, and the test was repeated. If both criteria were satisfied it was concluded that the data set could be expanded to $n = 402$ accurately by disjunctive cokriging, using the available simulations and the values of the co-variable.

Interpolation of S_0 values yielded a MVPE of 0.980 and a MSEP value of 0.541 (sample variance $s^2 = 1.588$). Quality criteria Q_1 and Q_2 were defined as Q_1: MVPE = 0.617 s^2 and Q_2: MSEP = 0.341 s^2. For scenarios S_1–S_4, 70 simulations were sufficient to satisfy both Q_1 and Q_2, a reduction of 80% relative to the 352 simulations of scenario S_0. Scenarios S_5 and S_6 needed 10 more simulations. A map of the conditional probability of exceeding the threshold leaching concentration of 25 mg nitrate/dm^3 at fertilizer levels corresponding to scenarios S_0 and S_6 is given in Fig. 1. Scenario S_0 resulted in leaching concentrations exceeding the criterion. The leaching criterion is scarcely exceeded in case of scenario S_6. In 1.4% of the field area the probability of exceeding the threshold level is higher than 0.05, implicating the area where the scenario is rejected.

We conclude that spatial interpolation is indispensable in making statements for areas of land from point observations. Efficient use of collected and simulated data may be helpful to overcome a large number of calculations. The gain is in particular due to a similar pattern of simulated yields appearing for successive scenarios.

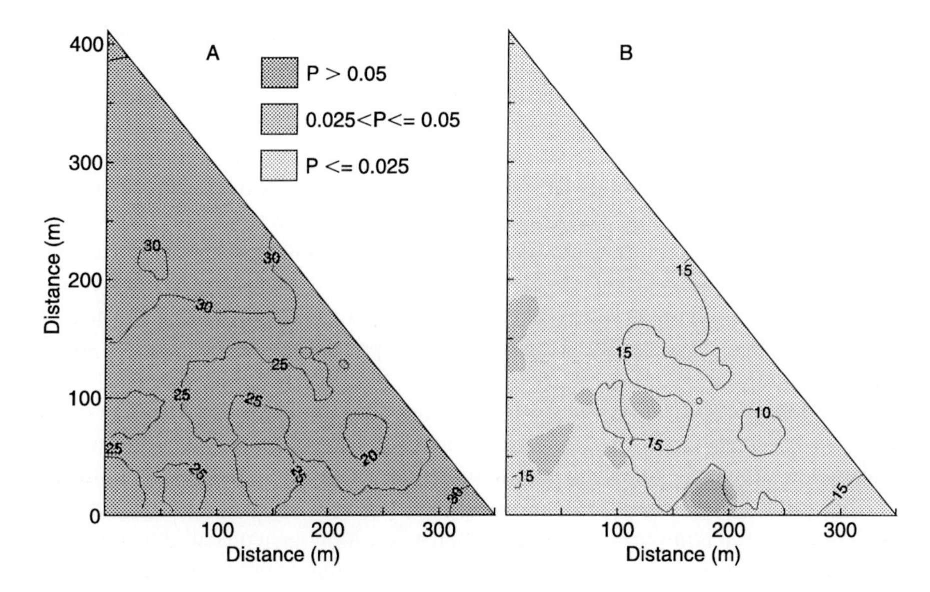

FIG. 1. Maps of simulated nitrate leaching concentrations (isolines) and of the probabilities of exceeding a 25 mg NO_3/dm^3 cut-off level (grey) with (A) scenario S_0 and (B) scenario S_6. From Finke & Stein (1994).

Pattern comparison

Similarities between preliminary yield mapping and field conditions may be useful for properly defining management strategies. Here we will consider geostatistical map comparison on interpolated grid maps. We will not consider comparison of patterns on chloropleth maps, which can usually be done with kappa statistics (Ripley 1996). An obvious way to compare patterns node by node is to calculate the correlation coefficient between the nodes on a map, or to regress the nodes of the grid on one map on the nodes of the grid of the second map. To compare patterns more generally, we propose the use of the correlation function. For two maps, M_A and M_B, consider a pair of variables $z_{ij,A}$ and $z_{i'j',B}$ measured at nodes (i,j) at M_A and (i',j') at M_B, respectively, separated by the distance $h = \sqrt{(i-i')^2 + (j-j')^2}$.

Mean values are m_A and m_B and variances are $s_A{}^2$ and $s_B{}^2$. Then $\rho_{AB}(h)$ as a function h is defined as:

$$\rho_{AB}(h) = \frac{E[z_{ij,A} \cdot z_{i'j',B}] - m_A \cdot m_B}{s_A \cdot s_B} \tag{1}$$

where E denotes the mathematical expectation. For similar patterns $\rho_{AB}(0) = 1$, whereas for opposite patterns $\rho_{AB}(0) = -1$. A decreasing cross-correllogram for increasing values of h denotes a decreasing correlation between $z_{ij,A}$ and $z_{i'j'B}$. If patterns show similarity for a distance larger than 0, $\rho_{AB}(h)$ displays a peak for that distance and gradually decreases towards 0. Totally unrelated patterns have $\rho_{AB}(h)$ equal to 0, whereas highly correlated patterns may have large $\rho_{AB}(h)$ values.

Space–time statistics

Space–time data are increasingly becoming available on agricultural phenomena. This is because of a growing awareness of the dynamic environment: sustainable agricultural systems require a continuing balance between inputs and outputs. In smart farming phenomena such as plant diseases, farmer's management in space and time, the distribution of leached pesticides in a homogeneous medium, or the changing patterns on successive satellite imagery, exhibit spatial patterns developing in the course of time.

Many differences exist between spatial and temporal data. The most important one is the cause–effect relation in time, and lack of it in space. Second, there is a scale difference. Spatial data are affected mainly by their representativeness, temporal data by both this and their periodicity. Finally, prior information may differ between spatial and temporal data. Stratification is important for spatial data, where homogeneous sub-areas provide essential basic information, whereas daily or seasonal fluctuations are important for temporal data. Deterministic modelling by means of agricultural models such as crop models, groundwater models and leaching models now commonly combines space and time data, e.g. to study the effects of different land use scenarios. In this paper we will present an overview of modelling space–time dependence.

Variables in space (S) and in time (T) are associated with their positions $s \in S$ and $t \in T$. It is common to model the variability in space and in time as a random field $Z(s,t)$ (Journel & Huijbregts 1978, Cressie 1991). Let the observation times be denoted by $t_j \in T$, $j = 1, \ldots, k$ and the spatial observation locations at t_j by $s_{ij} \in S$, $i = 1, \ldots, N_j$.

Modelling dependence in space and time

First we will consider independent spatial fields $Z(s,t_j)$ with a time-dependent expectation: $E_s[Z(s,t_j)] = \mu_j$. At each of these times, the spatial variability can be modelled by the variogram. Several spatial analyses can be carried out for each sampling day. This approach has been successfully applied to optimize sampling for groundwater management (Stein 1997).

As the next conceptual stage we consider replicates of spatial fields, i.e. fields with different expectations, but with the same variogram. The random field is $Z(s,t_j)$ observed at k different times, and $E[Z(s,t_j)] = \mu_j$. The variogram is defined as:

$$\gamma(h_s) = \tfrac{1}{2} E[\{Z(s + h_s, t) - Z(s,t)\}^2] \qquad (2)$$

where h_s is the distance in space and the expectation applies to each point in space for each time point separately. The variogram can be estimated using observations collected at each of the instances. An estimator of the variogram is given as:

$$\hat{\gamma}(h_s) = \frac{1}{2n(h_s)} \sum_{j=1}^{k} \sum_{i=1}^{n_j(h_s)} (z(s_i,t_j) - z(s_i + h_s,t_j))^2 \qquad (3)$$

where $z(s_i,t_j)$ and $z(s_i + h_s,t_j)$ have the same meaning as before and

$$n(h_s) = \sum_{j=1}^{k} n_j(h_s).$$

For each distance h_s each pair of points contributes to the estimate of the variogram, irrespective of the time of measurement. Such a model, although essentially including *spatial* dependence, does not take into account the obvious dependence *in time* that may exist between observations.

Example: wind erosion at a field-scale in Niger

In a recent study (Sterk & Stein 1997), we studied the effects of wind erosion. The variable we analysed was wind-blown mass transport during four successive storms. A field experiment was conducted in a plot of 40 m × 60 m in Niger in 1993. Twenty-one catchers were regularly distributed over the plot in three rows of six catchers, with the three remaining catchers placed between two rows.

Measurements for each location and storm were made with $k = 4$, and $N_1 = 20$ and $N_2 = N_3 = N_4 = 21$ (Table 2). The four storms were different in magnitude. The second and third storm were small storms, whereas the first and the fourth were heavy. Pooling of the four storms yielded a single data set with 83 observations for variogram estimation. The underlying assumption is that the four storms are independent temporal replicates of wind-blown mass transport, with similar spatial dependence structures. The storms were standardized to the mean mass transport value of all four storms ($m_0 = 75$ kg/m). The standardized observations were used to calculate variogram values. A spherical model was fitted through the estimated variogram, yielding a range of 52.1 m, a nugget value of 15.1 kg^2/m^2 and a sill value of 1220 kg^2/m^2. The

TABLE 2 Summary statistics and parameter values of the exponential variogram model of wind-blown mass transport for four storms on a farmer's field in Niger, 1993

	N	*mean* ($kg\,m^{-1}$)	*s* ($kg\,m^{-1}$)	CV	*Min* ($kg\,m^{-1}$)	*Max* ($kg\,m^{-1}$)	*Median* ($kg\,m^{-1}$)	C_0 ($kg^2\,m^{-2}$)	*A* ($kg^2\,m^{-2}$)	*b* (m)
Storm 1	20	102.7	36.9	35.9	24.0	213.6	102.4	28.3	2288	52.1
Storm 2	21	15.5	5.2	33.4	7.2	26.0	14.2	0.7	52	52.1
Storm 3	21	32.0	14.8	46.3	9.6	68.9	29.7	2.8	222	52.1
Storm 4	21	149.9	51.8	34.5	68.9	282.7	149.9	60.3	4874	52.1

standardized variogram was converted to four variograms $\gamma_j(h_S)$ where the range is the same for all storms, and storm-specific nugget and sill values depend upon storm-specific mass transport values. These variograms were used for interpolation and spatial simulation of the four storms. In this study, lack of spatial data was overcome by the additional assumption of constant spatial variability over time.

Similarly, we can define a random field for which the instances in time are replicates in space to model dependence in time. A random field therefore is observed at N locations, with different expectations but with the same variogram in time. The variogram is defined as:

$$\gamma(h_T) = \tfrac{1}{2}E[\{Z(s,t+h_T) - Z(s,t)\}^2] \tag{4}$$

where now h_T is the distance in time and the expectation applies to each point in time. The variogram can be estimated using observations collected at each location. An estimator of the variogram is then given as:

$$\hat{\gamma}(h_T) = \sum_{i=1}^{N} \frac{1}{2k_i(h_T)} \sum_{j=1}^{k_i(h_T)} (z(s_i,t_j + h_T) - z(s_i,,t_j))^2 \tag{5}$$

where $z(s_i,t_j)$ and $z(s_i,t_j + h_T)$ is a pair of observations at location s_j separated by a distance h_T, of which there are $k_i(h_T)$. For each distance h_T each pair of points contributes to the estimate of the variogram, irrespective of the location of measurement. Such a model, although it essentially includes dependence *in time*, does not take into account spatial dependence that may exist between observations.

A general spatiotemporal field can have many different forms. We will consider a field which obeys the intrinsic hypothesis: $E[Z(s+h_S,\ t+h_T) - Z(s,t)] = 0$ and $\mathrm{Var}[Z(s+h_S, t+h_T) - Z(s,t)]$ depends solely on $h = (h_S,h_T)'$ where h_S is the distance in space and h_T is the distance in time. Under the intrinsic hypothesis, the variogram $\gamma(h) = \gamma_{S,T}(h_S,h_T)$ in S and T is defined as:

$$\gamma_{S,T}(h_s,h_T) = \tfrac{1}{2}E[\{Z(s+h_S,t+h_T) - Z(s,t)\}^2] \tag{6}$$

To estimate the variogram from the observations, we first consider an isotropic model in space and time, i.e. the variability for each unit in time corresponds to the variability for each unit in space. The sample variogram as a function of r is computed by:

$$\hat{\gamma}(r) = \frac{1}{2N(r)} \sum_{j=1}^{N(h_T)} \sum_{i=1}^{N_t(h_S)} \{z(s_{ij} + h_s,t_j + h_T) - z(s_{ij},t_j)\}^2 \tag{7}$$

where $z(s_{ij},t_j)$ and $z(s_{ij} + h_s,t_j + h_T)$ is a pair of observations with a spatial distance at t_j approximately equal to h_S, the total number of such pairs being equal to $N_t(h_S)$, and a temporal distance approximately equal to h_T, the number of such pairs being equal to $N(h_T)$. The total number of pairs with separation distance equal to r equals $N(r)$.

Isotropy is seldom (if ever) encountered in the space–time domain, because no natural measure exists to combine spatial with temporal units. To account for this, the distance r is defined as $r = \sqrt{h_s^2 + \varphi . h_T^2}$, which includes the anisotropy parameter φ. This model can easily be extended to include spatial anisotropy as well. Such a model has the property that φ is constant in space and time, which may not hold in many agricultural studies where the variogram parameters are functions of time. For such situations we will return to a model with different variograms at different time points. Special cases are the spatial variogram $\gamma(h_S)$, which for a fixed time point depends only on h_S and has $h_T = 0$, and the temporal variogram $\gamma(h_T)$ which for a fixed location in space depends only h_T and has $h_S = 0$.

Prediction in space and time

To predict the values at unvisited locations in space or points in time we used kriging. If we consider the models with independent spatial fields or where spatial fields are replicates in time, predictions are only possible at the times of observation. Kriging weights can be derived only as a function of the distance in space, and hence observations at the same time point are required. For temporal fields as replicates in space only, predictions are possible at the N observation locations at different time points, because kriging weights can be derived only as a function of the distance in time.

Predictions can be carried out in two ways:

(1) Anisotropic space–time kriging using a space–time anisotropy parameter. This allows predictions to be made at every point in space and time, since the distances can be calculated and kriging can be applied.
(2) Two-step space–time kriging (2STK). This predictor is a linear combination of k predictors in space. In the first step, the predictors of the values at place s_0 and at times t_j, $j = 1, \ldots, k$, are determined. The predictors are linear in the observations. In the second step the value at location s_0 and time t_0 is predicted, using a linear combination of the predictors from step 1. The 2STK predictor explicitly uses spatial variograms changing in the course of time. As can be shown by standard methods, it is an *exact* predictor, i.e. at observation locations the observed values are themselves predicted and the kriging variance is equal to 0. The 2STK predictor has been compared for phytopathological data in this study with anisotropic space–time kriging (ASTK), using the anisotropic variogram and with temporal kriging (TK) assuming the intrinsic hypothesis to hold in time (Stein et al 1994).

Sampling in space and time: the experimental design revisited

The experimental design may yield crucial information on the effects of treatments and soil types on crop performance. We see as one of its problems that observations are collected at short distances from each other, hence maybe yielding too low variances, as observations close to each other are more similar than observations at a larger distance. Moreover, the calculated effects are considered to be representative for a catena, and hence need upscaling. We used geostatistical procedures to investigate this issue.

In a recent field trial experiment in Côte d'Ivoire the effects of three different fertilizer applications on rice and maize crops were estimated on five different soils (Hoosbeek et al 1997). A randomized block design was applied. Two procedures were defined to up-scale the information obtained from plots to catena/farm level. Procedure 1 involved the use of the soil classes to extrapolate the plot information to soil units at catena scale, the 'classical' method of assigning fertilizer recommendations to similar map units across an area. Procedure 2 involved the use of block kriging. In the initial project the plot layout was chosen to minimize the noise caused by spatial variability and to maximize the effects of the different treatments. In this design the soil is assumed to be homogeneous within the plots. As a result the plots were laid out in a cluster adjacent to each other for each soil type (Fig. 2a). The obtained results could be up-scaled with procedure 1. Up-scaling according to procedure 2 failed because the observations are adjacent and hence interpolation largely turns into extrapolation. The original design, used to obtain process-oriented information, ignores spatial variability and assumes homogeneous soil map units. We proposed an alternative sampling scheme (plot layout), in which we scatter plots each consisting of four smaller plots across the catena (Fig. 2b). Geostatistical simulations yielded data at these locations to which the average value for the soil type and that for the best treatment were added. The simulated yield data showed higher variances when compared with the measured yield data, due to the increased spatial variability. Such a design allows for interpolation between the plots and for kriging to up-scale to the catena level. Only the simulated spatially distributed data allowed the use of procedure 2.

Obviously, a mismatch existed between the use of the experimental design and requirements for spatial interpolation and up-scaling. Acknowledgement of spatial variability resulted in a different sampling design and in different up-scaling procedures.

Conclusions

From this study we conclude that spatial and temporal variability play an important role for smart farming at many scales. Several procedures are presented to analyse the variability and to use it for interpolation and other purposes. We conclude that geostatistical procedures are most valuable for *analysing* spatial and temporal variability and then to *use* it for practical purposes. Geostatistical procedures are useful to compare effects of land use scenarios. Modelling variability and interpolation in space and time may be applied for optimizing sampling schemes in space and in time and for making better use of expensive observations. Further, the procedures are flexible enough to include many forms of information that are already

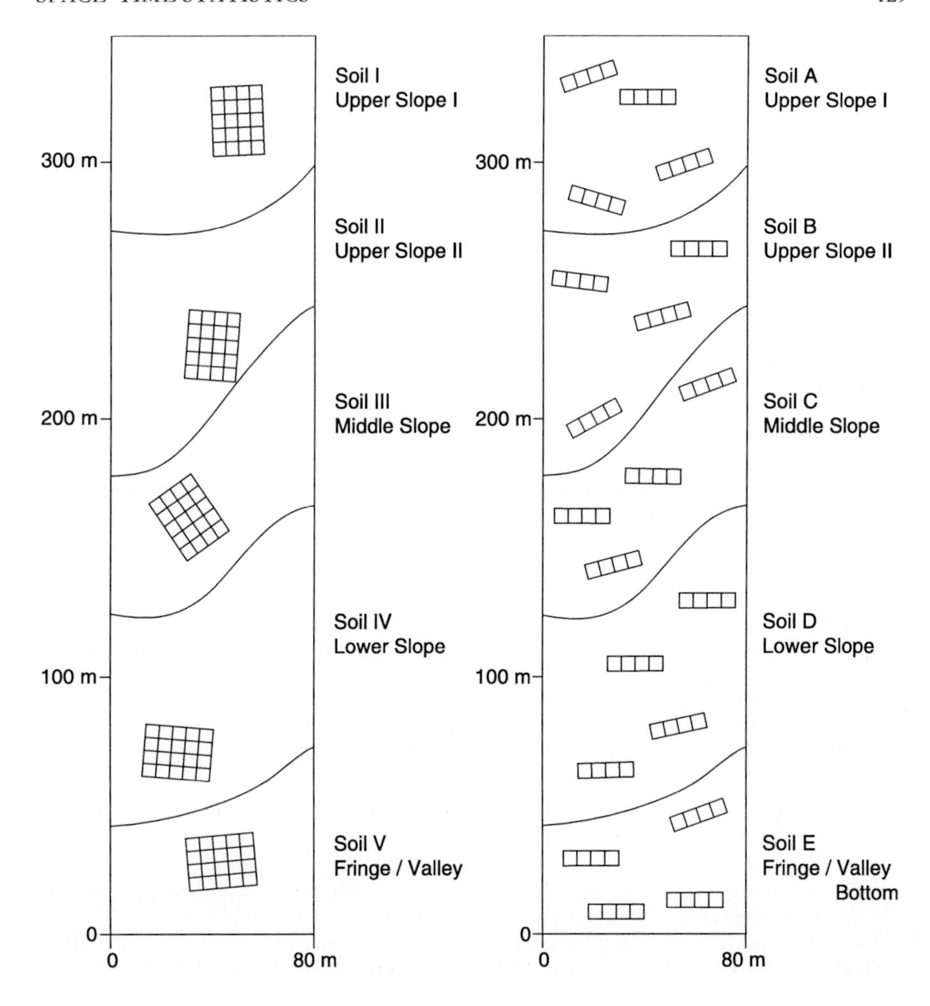

FIG. 2. Soil map of the catena with the five experimental plot fields, a suitable design for analysis by ANOVA (*left*) and proposed spatially randomized block design, maximizing analysis of treatment effects (per block) and allowing for interpolation (*right*)

available. As attention shifts towards smart farming with an emphasis on place and time dependence of management activity, additional statistical procedures are necessary. For example the experimental design may be better accommodated to meet the new challenges.

References

Burrough PA 1996 Environmental modelling with Geographic Information Systems. In: Stein A, de Vries Penning FWT, Schotman PJ (eds) Models in action. Research School of

Production Ecology, Wageningen, Netherlands (Quantitative Approaches in Systems Analysis 6)
Christakos G 1992 Random field modelling in the earth sciences. Academic Press, San Diego, CA
Cressie NAC 1991 Statistics for spatial data. Wiley, New York
Finke PA, Stein A 1994 Application of disjunctive cokriging to compare fertilizer scenarios on a field scale. Geoderma 62:247–264
Hoosbeek MR, Stein A, van Reuler H, Janssen BH 1997 Spatial and temporal variability of agronomic data for prediction at different scales. Geoderma, in press
Hutson JL, Wagenet RJ 1991 Simulating nitrogen dynamics in soils using a deterministic model. Soil Use Manag 7:74–78
Journel AG, Huijbregts CJ 1978 Mining geostatistics. Academic Press, New York
Matheron G 1976 A simple substitute for conditional expectation: the disjunctive kriging. Proceedings, NATO ASI: Advanced geostatistics in the mining industry. Reidel, Dordrecht, p 221–236
Mulla DJ 1993 Mapping and managing spatial patterns in soil fertility and crop yield. In: Robert PC, Rust RH, Larson WE (eds) Soil specific crop management. American Society of Agronomy, Madison, WI, p 15–26
Ripley BD 1996 Pattern recognition and neural networks. Cambridge University Press, Cambridge
Stein A 1997 Monitoring tools and observation density: spatial and temporal variability. In: Van Lanen HAJ, Simmers I (eds) Monitoring for groundwater management in (semi-)arid regions. UNESCO, Paris, in press
Stein A, Kocks CG, Zadoks JC, Frinking HD, Ruissen MA, Myers DE 1994 A geostatistical analysis of the spatio-temporal development of downy mildew epidemics in cabbage. Phytopathol 84:1227–1239
Stein A, Brouwer J, Bouma J 1997 Methods for comparing spatial variability patterns of millet yield and soil data. Soil Sci Soc Am J 61:861–870
Sterk G, Stein A 1997 Mapping wind-blown mass transport by modeling variability in space and time. Soil Sci Soc Am J 61:232–239
Van Uffelen CGR, Verhagen J, Bouma J 1997 Comparison of simulated crop yield patterns for site-specific management. Agric Syst 54:207–222
Wagenet RJ, Hutson JL 1989 Leaching Estimation and Chemistry Model: a process-based model of water and solute movement, transformations, plant uptake and chemical reactions in the unsaturated zone. Continuum Water Resources Institute, Cornell University, NY
Webster R, Oliver MA 1989 Optimal interpolation and isarithmic mapping of soil properties. VI. Disjunctive kriging and mapping the conditional probability. J Soil Sci 40:497–512

DISCUSSION

Barnett: I was very interested to hear your comments on temporal and spatial interconnections, and particularly the earlier remarks you made about use of kriging for spatial study. Later on you seemed to step back and said 'Let's look at time, here it is for three years'. I am concerned that we are not developing genuinely integrated spatial–temporal models, and I would like to make a small suggestion. If you think of ordinary time series analysis in the spectral domain, the spectrum is the determinant of the behaviour of the process. Recently, some efforts have been made to extend the notions of the spectrum to spatial processes: looking at a two-dimensional spectrum which is a representation of a spatial process. It seems to me that once you recognize that this might be useful in the spatial domain (even if directionality does not apply here as it does in a temporal domain) then there seems to be no reason why you should not define a three-dimensional spectrum where the three components are x, y and t— namely the spatial dimensions and the time dimension. It is going to be very difficult to use this approach, to interpret it, and so on. I have seen primitive movements in this direction in areas like medical diagnostics, particularly in using encephalograms, for instance. I do think there could be some future for agricultural-type problems.

I also have a brief comment about design. I'm a little worried about you changing your design in the way you did. In the first design, if you think of the concept of eastings and northings, you have rows of columns which represent eastings and northings which are crossed with each other. Once you pull them apart you lose that crossing and it becomes nested. So if there is information about relative easting positions, I think you might have difficulty in trying to sufficiently estimate those in what is in other respects a more representative design.

Stein: The drawback of using spectral analysis in three dimensions could be that it doesn't properly include the cause–effect relationships in time. Another drawback may be that until now you were expected to have a grid sampling with equal distances in space and/or in time to be able to properly estimate the Fourier transform.

The other point that you mentioned concerned the question of losing the crossing effect: I agree with you on this. This was our first attempt. This type of design might not be as appropriate as it could have been, but I think this could be one alternative to make it more appropriate for the specific challenges we have to meet.

Rasch: I have also been thinking about the problem of taking time as a third or fourth component in spatial statistics. Technically this is feasible, but it could be tricky, because time has another behaviour in addition to spatial variation. This depends, of course, on the measurement points: for example, if you take measurements during an 11 d period it will not matter too much if you don't fertilize in the mean time, but if you look over a whole year there will be seasonal components, which you will not have in the space. Perhaps it is also possible to make the time series trend-free and free of these seasonal components, then we can use this three-dimensional approach and later on, in the predictions, add all the seasonal effects to make a better prediction. Otherwise it will become too complicated.

Stein: I basically agree with your standpoint. Separating the obvious major effects from local, small scale disturbances is generally OK. There is a warning in the literature (Matheron 1971), though, that the spatial and temporal dependence structure is totally different for these residuals than for the random field as a whole. But the aim may not be to characterize the field spatially (and temporally), but rather to make interpolations or to simulate. An interesting approach is also given by Cressie (1997). He coined the idea of combining a standard time series model, such as a first order autoregressive (AR[1]) model, with a spatial model. This would have the benefit of having a cause–effect relation in space, governed by this AR(1) model with a spatial model that describes the local and the global variation. It was at that time, however, unclear whether such a model would be fully equipped to address a broad range of practical issues.

Rasch: In a factorial experimental design one tries to put all the plots as close together as possible, because we work in blocks. Blocks are a method for reducing the influence of soil variability. If you want to estimate the soil variability, you have another aim, and then you should distribute your observations over the area where you would like to have information. I would never combine a factorial design (estimating the effects of factor combination) with investigation of spatial variation. You should do one or the other, but not both together.

Stein: I hope there will be opportunities to combine both. We need a combination of estimating effects on the one hand and scaling or making statements for larger areas of land on the other. I don't know whether the way we have done this is appropriate; it is just an addition to the idea of experimental design.

Rasch: Then we have a series of observations of different locations that can give better information about the factor × location interaction.

Braud: I have a question about your first example dealing with erosion data. The data presented in your paper indicate that the mean is related to the variance: when the mean is low, so is the variance, and vice versa. Did you take that into account when you calculated the value of the average variogram?

Stein: That is a good point. We standardized the data, that is we correlated for the differences in variance between the fields. As such, the different variograms were fully compatible, were pooled, and a single model was derived. This model was then back-transformed using the differences in observed variance. It sounds like an interesting idea to make a weighted combination with weights proportional to the variance. But that may be something to study later.

Braud: This is something commonly seen with regard to rainfall data. One way to cope with this relationship between mean and variance is to divide each field by its spatial standard deviation before calculating the average variogram.

Stein: One thing we did recently to follow up the study I presented, was to compare our results with the individual variograms. That is, we took the 21 data and estimated for variograms. When you do this you realize that some inherent variability may still be present. This is probably also related to your question about differences in variance.

Braud: I suppose that if you do have different sills for the individual variograms, it is because the variances of your individual fields are different.

Stein: But as I showed, we have standardized the data to make one common model, which we modified later to obtain a specific model for each storm. This model is then related to the proper variance by having a similar sill value.

Gómez-Hernández: I would like to comment on the suggestion of using a three-dimensional spectrum to handle spatiotemporal phenomena. Either to describe the spatial variability or spatiotemporal variability, there is a perfect correspondence between covariance and spectrum. There is nothing to be gained by going to the spectrum domain. Besides, as soon as you jump into the spectrum domain and start working with frequencies you lose contact with reality. If anything can be done with spectrum it could also be done by using covariances. You can account for temporal correlation with the same flexibility. However, the temporal component has a different behaviour from the spatial variability and it may be difficult to put them all together in a single covariance model.

Thompson: When you first listed the differences between the spatial and the temporal settings, the second one was scale representativeness versus periodicity. Could you elaborate on this? It seems certain that you would see periodicity in a field where there are lines where the farmers fertilized and planted in the past.

Stein: Perhaps this is simply a matter of wording. When you talk about spatial representativeness, you are talking about whether this point is representative of large areas. If you talk about variability in time, periodicity is the major topic to deal with, although this could be just a different approach to the same topic.

Mulla: Several years ago I published a paper in which I compared two-dimensional spectral analysis with two-dimensional semivariogram analysis (Mulla 1988). I found that the spectral analysis is quite powerful when it comes to elucidating the periodic behaviour in a random variable, but the semivariogram analysis is powerful when it comes to interpreting short range structure that is non-periodic. So there are advantages and disadvantages in both methods. Perhaps for the time series analysis a little of both would be of interest. Maybe there are differences in the power of the spectral analysis versus the covariance structural analysis.

McBratney: If we're talking about these techniques we might as well be up to date and ask why we don't use three-dimensional wavelet analysis to deal with non-stationary aspects of these problems. The real problem is how we are going to deal with the non-stationarity in time.

Barnett: My suggestion was for stationary processes, but I agree with you entirely.

References

Cressie N 1997 Aggregation and interaction issues in statistical modelling of spatio-temporal processes. Geoderma, in press

Matheron G 1971 The theory of regionalized variables and its applications. Technical Report no. 5, Les Cahiers du Centre de Morphologie Mathématique, Fontainebleau

Mulla DJ 1988 Using geostatistics and spectral analysis to study spatial patterns in topography of southeastern Washington State, USA. Earth Surface Processes & Landforms 13:389–405

General discussion II

Rabbinge: Underlying our discussions this morning was the premise that smart farming is an agricultural method that is going to develop further. We are no longer discussing, as we did yesterday, whether or not it is a useful exercise: we have accepted the broad definition of precision agriculture.

McBratney: Could you clarify whether the terms 'smart farming' and 'precision agriculture' are synonymous?

Rabbinge: At one level, they are exactly the same. But if you want to be pedantic, precision agriculture is related only to the fine-tuning of inputs in time and space, whereas smart farming encompasses not only yield-limiting factors such as water and nutrients but also the yield-reducing factors, and fine-tunes in time and space structural changes and activities during the growing season. In fact, smart farming could be considered to be modern high-knowledge intensive agriculture, making use of a whole toolbox of insights (from basic science to applied science) to maximize productivity while minimizing negative environmental impact.

What we are now discussing are some of the tools. If you use the narrow definition of precision agriculture, then we need three different types of tool: the models, the sampling and the maps.

From Vic Barnett's contribution it was clear that if you use mechanical and explanatory models as a black box for predictive purposes, you may end up with totally wrong results (Barnett et al 1997, this volume). That is why I addressed the importance of using the right tool for a specific situation. Then we are using models correctly. We probably need a mixture of explanatory models, summary-type models, empirical models and regression models. An optimal combination may best fulfil our aims.

Groffman: I don't have any argument with the premise: certainly, we're going to go ahead with developing precision agriculture. We have repeatedly stated that the motivation for this is increased efficiency, improved profitability and improved environmental performance. However, we have seen no data on those things. We have also not addressed the old problem of conflicts between improved productivity and environmental quality. For example, in David Mulla's paper we saw an advance that leads us to apply more phosphorus fertilizer to potentially highly erosive soils (Mulla 1997, this volume).

In other words, precision farming is obviously something that is going ahead, but if we want farmers to adopt these practices, what data are we presenting to them to say that there is improved environmental performance and increased efficiency without conflicts between the two? I'm certainly willing to move ahead and discuss tools, but we should establish this first. There's a very useful line in Johan Bouma's abstract where he says 'in fact, we are faced with a very strong commercial technology push which is not accompanied by the build-up of adequate agroecological expertise.' Certainly, we have talked a lot about tools — technically, precision management is feasible — but what are we getting from it?

Bouma: Pierre Robert mentioned that some farmers are voting with their feet: 80% of all the sugar beet farmers in Minnesota are applying precision farming because they gain US$40–60 per acre. I agree with Peter Groffman that we need more specific cases to illustrate the benefits of precision farming. We have also done studies on balancing production and environmental requirements. Sometimes you have to sacrifice a little bit of production because of environmental thresholds. If you can show farmers, because there has been such an overuse of fertilizers, that they can reduce this input considerably without losing yield and at the same time they can improve environmental quality, you are in business. As scientists, we should package these examples as case studies showing the advantage of precision management over traditional management.

McBratney: We often talk about these twin goals of efficient production and lowering environmental risk. Obviously, for the farmer, the ultimate goal has to be profitability. I have begun a project with some economists in which we are trying to come up with a function that will tell you the profit you make given your inputs, but we also have to put the environmental cost into the equation. Thus, the real profit is your gross margin, plus or minus the environmental cost. The problem at the moment is that society hasn't really worked out how we're going to measure environmental cost. There are various ways we can do it, but that to me is fundamental to getting away from many of these arguments. If we could put the environmental cost into the equation — the external costs — then I think we can show more readily that the precision farming concept is better than the conventional one. I think this a better approach than just having environmental cut-offs and guidelines and so on: you actually have to think about what it is going to cost to fix up the problem that you're creating. If you apply too much fertilizer, there's a missed opportunity cost because you're wasting fertilizer, but you're also causing a environmental problem that has to be fixed. These components have to be costed and put into the equation and, then, if there is a cost to be paid it has to be borne by the farmer. Then profitability is clear.

Rabbinge: That seems very reasonable, but how are you going to quantify in financial terms a nitrate emission of 50 mg/l, for example? Who is going to decide this? This is why I think we should specify the aims, such as environmental thresholds, in their own dimension. This technology will help us develop, in a preventive sense, cropping management systems which will lead to high productivity and high environmental efficiency.

McBratney: All I'm saying is that by doing what I propose the twin goals are brought onto the same scale: the scale of the mighty dollar!

Rabbinge: But that's the problem: there is a normative element in it and that requires societal discussion. That is more than science can deliver and it explains why economists have such difficulty with this issue. For 20 years we have been discussing the concept of the 'green gross national product'. In the early 1970s people suggested that we 'green' the gross national product to make clear that we should take better account of non-renewable resources, and that this should be translated as a cost in the national product. I've been involved in these discussions, and the problem is that it has been a dead end because of these normative elements: if there are normative elements involved this is then a decision which has to be made in society and by policy makers.

Barnett: I was a little concerned that you gave the impression that no one is attempting to put an economic value on the environmental dimension. In fact many people are doing it, unfortunately in different ways in different countries and sometimes on rather strange principles—often on the bandwagon of the 'sustainability' argument. I just wanted to set a slight counterbalance and say that we should be encouraging people properly to cost out the environmental influence and effect, and not to give the impression that this is not worth doing.

Rabbinge: I'm saying that the costs should be made explicit in their own dimensions. Then it is possible in negotiations and policy making to see the trade-offs. Such trade-offs are in their own dimension and that helps to make a final decision. You have to keep the whole process open and transparent. I've been involved in 'sustainability' discussions and it seems that there are hundreds of different definitions of the term, so it doesn't make much sense to say that sustainability is the ultimate aim.

Groffman: This issue is well beyond the scope of our discussions. All I was asking for were some data, for instance showing that nitrate concentrations are lowered by precision management. We need some figures to show that these things work, particularly in terms of environmental quality but also in terms of profitability. As you pointed out, the environmental costing is very complex. We would need to factor in the economic context in terms of agricultural policy. Sugar beet is a good example of this need. This is a highly subsidized crop in the USA, and this should be considered in the economic and environmental analysis.

Kropff: Indeed, we need figures to show that precision agriculture works with respect to profitability and environmental quality. However, it is very important, as a baseline, to understand the background of spatial variability in yield, so that the problem can be solved by management. Models are crucial in this context. I thought that Vic Barnett's paper this morning was significant: if you go for site-specific management you have to understand why there are differences in yield, because otherwise you can't eliminate them (Barnett et al 1997, this volume). Vic suggested that with the current models we cannot explain differences in yield between sites. In the GCTE (Global Change and Terrestrial Ecosystems) project (a core project of the International Geosphere–Biosphere Programme; IGBP) we have several networks that focus on the evaluation of models. We started with wheat models. Twelve different modellers met in The Netherlands and simulated yield potential without pests, diseases and weeds, with adequate nitrogen and no water stress for two specific environments. These models ranged from very simple models to very complex ones that can only be run on a Cray supercomputer. There was a tremendous variability in simulated yield potential with the same weather data. The group is now in the process of trying to explain this. The main reason is probably that these models have been developed with different objectives, and they were calibrated to specific environments. For their original purposes these models were obviously quite functional, but not all models are of use for predicting yield in a broad range of environments. This does not mean that these models are useless; on the contrary, they can be very helpful in explaining yield differences as a result of different factors, but they cannot be used in any situation without knowledge of the contents and the objectives of the model.

We subsequently looked at models with rice. This was interesting, because rice is a simpler system in that it is irrigated. We conducted a comparison study with four modellers from Japan, Australia and the Philippines (Kropff et al 1995). We all had been doing experiments to achieve yield potential with different genotypes in our own environment. In Australia they got 15 tonnes/ha experimentally, in Japan 7 tonnes/ha and in the Philippines 6 tonnes/ha in the wet season rising to 10 tonnes/ha in the dry season. So there was enormous variability in yield. All these models were quite good at predicting straight away this enormous range in yield potential on the basis of differences in radiation and temperature, because they were developed to simulate yield potential and because all modellers based their models on parameter sets that focused on achieving yield potential. I really think that further studies to test the crop models for broad ranges of environments are crucial. Therefore we have to start combining ecophysiological models and statistical techniques to understand genotype/environment interactions, so that we can use the models as an analytical tool to analyse data from multilocation experiments. The combination of statistics and simple mechanistic models is the key for the future of precision management.

Barnett: I find those remarks very interesting. I am encouraged to hear that at least in one area, namely rice modelling, we seem to be getting some useful results at the yield potential level. But the argument that I have had made to me when I talk about the fact that these wheat models do not seem to predict actual yields, is that we are looking at them the wrong way: 'they're not designed to predict yield, they're designed to have much more general use to predict yield potential'. But now we are hearing they do not predict yield potential either because they vary enormously one from another under apparently similar circumstances. Then you tell us that there are other valuable uses for them. Could you give me an example of this? What is the value of the model beyond it being able to predict either actual yield or potential yield?

Kropff: There are many ways in which these models can be useful. Indeed, the rice models study showed that our current knowledge facilitates the prediction of yield potential in a wide range of environments, *if* it is properly parameterized and not calibrated to simulate lower actual yields without introducing the proper mechanism. When models are just calibrated by adjusting parameters to fit model output to data, the models lose their value, of course. In many cases, when these models are calibrated for a given environment, for specific management options they predict reasonably well, as has been shown in many scientific papers. The problems occur when you apply a model to an environment completely different from that which it was designed for, because these models don't include all the factors. If they were to include them all, no one would be able to run them, because we don't have all the data on soil characteristics, pests and diseases, and so on.

First you have to identify the key factors explaining yield differences among sites, then you have to model the effects of these factors properly, test the model and try to design management options to improve the system. Other ways in which models can be useful are related to the integration of knowledge in basic disciplines, setting priorities for research in these basic areas and also in analysing backgrounds of yield difference in experimental treatments, next to of course predicting yield potential and water- and nitrogen-limited yields.

Barnett: I get very nervous when the word 'calibration' is used in this context, because the only examples I have ever seen of so-called calibration has been in post hoc exercises where it has been observed that the model does not in fact predict what is happening, so you sort of 'tweak it around' until it begins to match up in some way. What you've then done is 'calibrate' it. But it has not been calibrated, you have just changed the initial conditions, so that on that particular set of data it happens to work. Now I am genuinely hoping to see examples where models have broader application than that, even as management tools (indeed, especially as management tools, because I do not see them as predictive tools; they clearly do not work as such). So much time and effort has been spent on the models, and so much subtle and clever science has gone into the development of them, that we must indeed hope ultimately to find some value in them.

Rabbinge: Models can be of value in many respects. They can be predictive, descriptive and integrative. They can act as a means of communication between specialists (say physiologists on the one hand and economists on the other) and they can be used for explorative purposes. They also have a heuristic value in continually improving knowledge and insight. Indeed, for each purpose, different models are needed. But the claims of many of the modellers are that they often have, in addition to a predictive value, an explorative value. It is important here that you distinguish clearly between the different values and that you also make clear that models should not be used in the wrong way. It is not the models that create the problems, but their misuse, especially in the predictive area. I think all of us agree that if you have a model which is used outside its field of validity it will give wrong results. Because of this we should be reluctant to use models as 'black boxes'.

Mulla: One of the most interesting applications of crop models is to predict what would happen to production of grain in the event of changes in precipitation, CO_2 content or temperature due to global warming. I'm totally unaware of how accurate these predictions are, and I'm quite sceptical in view of what you have talked about in terms of how well those models work for your data set. There is enough uncertainty about global change and what it might do, and the models themselves seem to be somewhat uncertain as well.

Barnett: It is fascinating you say that because that was the initial stimulus for our work: a recognition that people were beginning to use these models not only in the general potential yield prediction sense but also in the global warming debate. A number of us were so concerned that there were many uncertainties in that field, that we should try at least to tie down one of them — do these models bear any relationship in their output to what happens in the real life situation? We looked at it precisely for that reason and became very concerned at the end in finding that they do not. It is not enough to say that the global warming situation is going to be rather different. You have at least to find some sort of rough predictive pattern that is going on if you are to have any faith at all in using the models for prediction in the global warming debate.

Rabbinge: In an earlier Ciba Foundation Symposium on *Environmental change and human health* (Ciba Foundation 1993), I made the point that other effects due to economic measures are going to affect the outcomes of yields much more significantly than the small changes due to climate change.

McBratney: Just to try and rationalize why the models don't work, precision agriculture is local management based on local information, and I expect that the local interactions are going to be different everywhere. Therefore, I don't expect a generic model to be predictive. Many of the decisions that are made in precision agriculture assume that you have the same response function everywhere, but you don't.

Rabbinge: In the generic models the basic relations are always the same, but the input factors differ widely.

Gómez-Hernández: What we have been lacking so far is discussion on the uncertainty of the input data and how this uncertainty influences predictions. A comment was made earlier about the need for confidence intervals at the entrance of the model. There is also a need for presenting confidence intervals of the model predictions. I believe that some of the poor behaviour of the models described in Vic Barnett's paper (Barnett et al 1997, this volume) could be explained by the large uncertainty in the input parameters. Vic mentioned that some sensitivity analyses had been done with these models; my question then is, was the range of model responses in the sensitivity analyses the same over the whole set of models? Another question would be, did you check the consistency of the models, that is, comparing the models responses for different models that used the same input parameters?

Voltz: I agree. In doing model validation we often compare simulated and observed values, but in fact the differences between these values arise not only from errors in the model concepts, but also from errors in parameter values, in model input data and in the observations used for validation. For example, when you said that your statistical model only explained 25%, in fact you couldn't hope to explain the whole lot because the observed yield had an uncertainty, and also there was a large uncertainty in the input data. If you look at the propagation of these errors, you could evaluate the variance on the output and take that away from the whole variance. Thus instead of looking at the ratio between the variance of the errors and that of all of the observation, you could subtract it, in which case we might have a more optimistic view of the predictive power of the model.

Barnett: I welcome many of these comments. With regard to the first question you addressed, we did follow all aspects of that.

The second point that you made is a very important one. 'OK, so the models don't predict well, but are they at least consistent with each other?' Well, no, as I explained in my paper, absolutely not. There is a correlation of only about 0.2 between the values given by any one model and any other, and this is in some ways the worrying thing about them — it makes you question the very science that is underpinning these models.

Bouma: The proof of the pudding is in the eating. The conventional system of agriculture where fields are uniformly managed is our point of reference. Precision agriculture should result in the fact that some features of your agroecosystem must function better than in the conventional system. Models are one tool for defining such alternatives. What we need is integrated case studies, where one considers the entire system. For this, we have to work with economists: we cannot work without them because we have to consider economic and environmental issues and the entire operation is a matter of profitability. We are getting into a phase where there are already farms that have been generating yield maps for several years. Using these databases we can look at the entire production system, compare precision with conventional management, and show the benefit. In our excitement about research it is sometimes easy to lose the whole picture and to focus only on small aspects.

References

Barnett V, Landau S, Colls JJ, Craigon J, Mitchell RAC, Payne RW 1997 Predicting wheat yields: the search for valid and precise models. In: Precision agriculture: spatial and temporal variability of environmental quality. Wiley, Chichester (Ciba Found Symp 210) p 79–99

Ciba Foundation 1993 Environmental change and human health. Wiley, Chichester (Ciba Found Symp 175)

Kropff MJ, Williams RL, Horie T et al 1995 Predicting the yield potential of rice in different environments. In: Humphreys E, Murray EA, Clampett WS, Lewinn LG (eds) Temperate rice achievements and potential. Proceedings of the Temperate Rice Conference, 1994. Yanco, Australia, p 657–664

Mulla DJ 1997 Geostatistics, remote sensing and precision farming. In: Precision agriculture: spatial and temporal variability of environmental quality. Wiley, Chichester (Ciba Found Symp 210) p 100–119

Variability and uncertainty in spatial, temporal and spatiotemporal crop-yield and related data

Alex. B. McBratney, Brett M. Whelan and Tamara M. Shatar

Australian Centre for Precision Agriculture, Department of Agricultural Chemistry & Soil Science, University of Sydney, Sydney, NSW 2006, Australia

Abstract. Application of the theories of precision agriculture to the practicalities of broad-acre farming relies on successful handling of the ramifications of uncertainty in information, i.e. information pertaining to the spatial and temporal variation of those factors which determine yield components and/or environmental losses. This paper discusses the uncertainty of yield and related variables as measured by their spatial and temporal variance. The magnitude of these two components gives a suggestion as to the appropriate scale of management. Simultaneous reporting on spatial and temporal variation is rare and the theory of these types of process is still in its infancy. Some brief theory is presented, followed by several examples from the Rothamsted classic experiments, yield-monitoring experiments in Australia, a long-term barley trial in Denmark, and a soil moisture monitoring network. It is clear that annual temporal variation is much larger than the spatial variation within single fields. This leads to the conclusion that if precision agriculture is to have a sound scientific basis and ultimately a practical outcome then the null hypothesis that still remains to be seriously researched is: 'given the large temporal variation in yields relative to the scale of a single field, then the optimal risk aversion strategy is uniform management.'

1997 Precision agriculture: spatial and temporal variability of environmental quality. Wiley, Chichester (Ciba Foundation Symposium 210) p 141–160

A successful implementation of precision agriculture will be dependent on the ability of individual growers to manage their crops differentially to achieve the twin goals of maximizing yield or profit while simultaneously minimizing environmental impact. The major obstacle to this is uncertainty of information, i.e. information pertaining to the spatial and temporal variation of those factors which determine yield components and environmental losses.

This paper will not focus on the determinants of yield but rather on the uncertainty of yield and related variables as measured by their spatial and temporal variance. The magnitude of these two components gives a suggestion as to the appropriate scale of management. We will proceed by discussing the amount of spatial variability that has been reported in the literature, followed by a discussion of some recent results on purely temporal variation. It has been quite rare to report simultaneously on spatial and temporal variation and the theory of these types of process is still in its infancy. Some brief theory is presented, followed by several examples.

Quantifying spatial variability

The spatial variation in crop yield at regional scales is often considered to be the consequence of variability in the interaction between crop genetics and environmental factors (Bresler et al 1981). However, at the field scale, site-specific variation in soil type, nutrient levels, soil moisture content and structural integrity will significantly contribute to the spatial variability in crop yield.

Much of the early work on spatial variability of crop yields using uniformity trials was reported in a remarkable paper by Smith (1938), which remains unsurpassed to this day. In fact, he presented one of the earliest yield maps, derived from data collected in Australia during December 1934 (Fig. 1).

It is typical of modern yield maps in that it shows approximately 100% variation in yield from lowest to highest across the area. For data such as these, and from many other studies that had been reported earlier, he showed that the variance of crop yield within an area could be described by a power law with parameter b (a coefficient of yield uniformity where the smaller b the more uniform the crop; Table 1).

$$V(A) = kA^{b} \tag{1}$$

Much later, geostatisticians (Journel & Huijbregts 1978) studying ore variation suggested a model of the type

$$Y(h) = ch^{\theta}$$

If we infer this for yield, then

$$V(h) = ch^{\theta} \tag{2}$$

These models have an infinite variance — in fact, the variance increases similarly for all scales. This is in fact a fractal model (Mandelbrot 1977). It infers a spatial or temporal process with a surface fractal dimension $D = (4 - \theta)/2$ if the process is one dimensional (Berry & Lewis 1980, Burrough 1981). The remarkable similarity between (1) and (2) suggests that for crop yields we can postulate $\theta = 2b$ and $b = 2 - D$. So we can characterize the variation using either b, θ or D. Smith's model is really a fractal model of yield variation expounded some 40–50 years before these came to vogue.

More recently, Bresler et al (1982) have shown coefficients of variation (CV) of 11–20% for irrigated corn and 12–26% for irrigated winter wheat. They correctly contend that the sample and overall field size will affect the mean and variance observed if any spatial structure exists in the yield data. Miller et al (1988) took $1\ m^2$ wheat yield

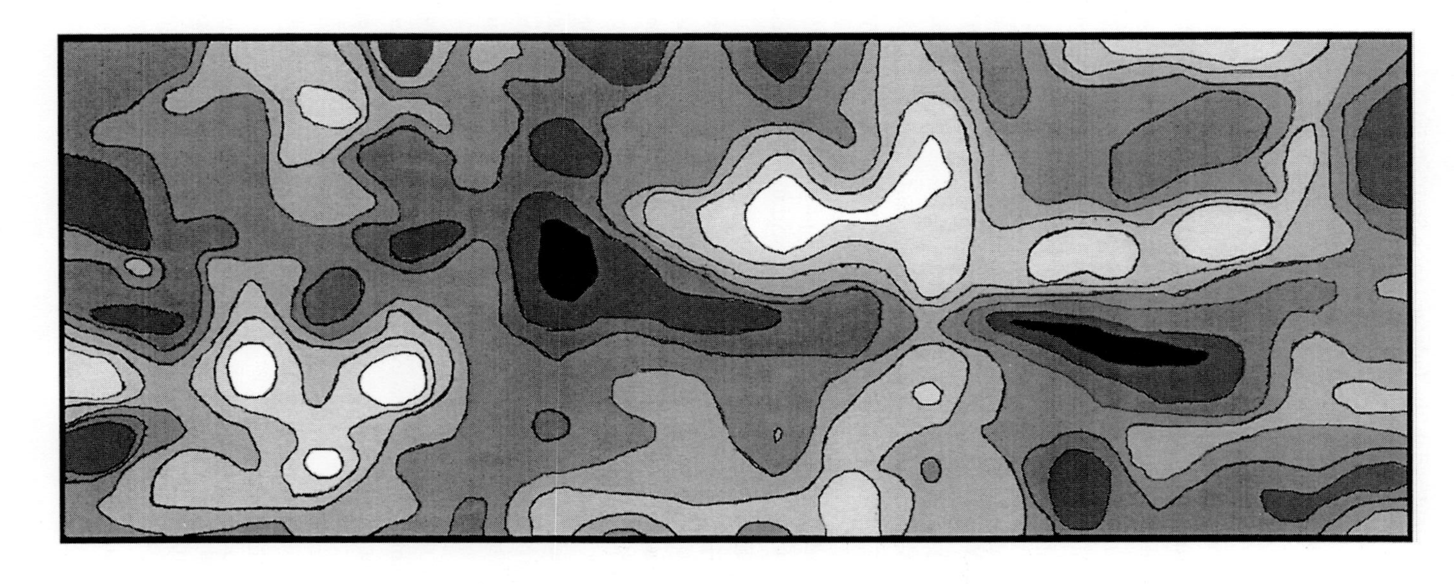

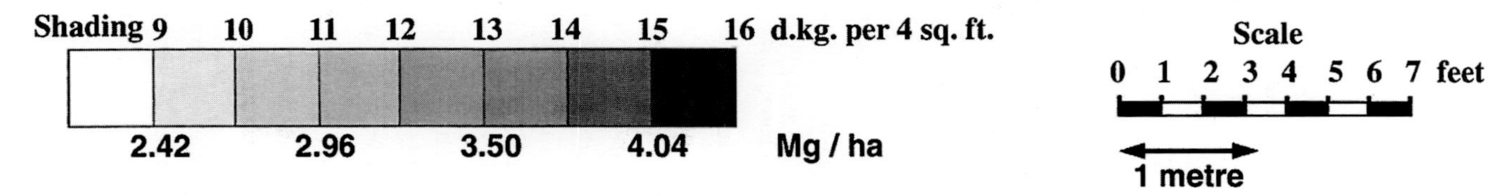

FIG. 1. Wheat yield map derived from data collected in Canberra during 1934 and published in Smith (1938).

TABLE 1 Within-field crop yield variability (adapted from Smith 1938)

Crop	*Year*	*Location*	*Plot size (m^2)*	*Number of plots*	*Mean yield (t/ha)*	*CV for 0.01 ha (%)*	*b*	*D*
Wheat	1911	Rothamsted	8.1	500	2.2	6.3	0.37	1.63
	1932	Rothamsted	0.08	1092	2.5	4.4	0.51	1.49
	1913	Nebraska	2.8	224	2.4	4.9	0.49	1.51
	1920	Missouri	0.3	3100	1.4	3.7	0.80	1.20
	1920	Missouri	0.3	3100	1.4	6.9	0.47	1.53
	1938	Australia	0.05	1080	3.2	1.7	0.72	1.28
	1938	Australia	0.2	54	2.7	3.1	0.44	1.56
Irrigated wheat	1935	Idaho	1.4	1440	4.2	10.5	0.08	1.92
Potatoes	1924	West Virginia	3.3	186	14.6	10.5	0.37	1.63
		West Virginia	3.3	290	10.5	18.9	0.18	1.82
		West Virginia	3.3	3309	7.1	25.1	0.26	1.74

samples on a 20 m × 50 m grid over 10 ha and reported a CV of 27%. Carr et al (1991) found the within-field yield CV for wheat (10 m^2 samples within 0.25 ha), harvested according to soil type, to range between 7% and 37% in one year.

Guitjens (1992) sampled irrigated winter wheat in 1.6 m^2 continuous plots along two transect lengths and reported CVs of 15–40% and 27–36%. He concluded that as water deficit increased and yield correspondingly declined, the CV increased linearly. The results may be somewhat ambiguous due to the use of two transect lengths. Somewhat more rigorously, Hunsaker (1992) sampled sixteen 12.2 m^2 plots of irrigated cotton within each of twelve 0.35 ha basins under three irrigation treatments. In two successive seasons, CVs for high-to-low irrigation treatments ranged from 7–17% and 12–26%, displaying a significant decrease in CV as irrigation (and yield) increased.

That spatial variability in crop yield occurs within fields and its magnitude varies between fields is not a contentious issue. However, the significant local influences on spatial yield variation and the relationship between mean yield levels and CV remain unclear.

Quantifying temporal variability

A fractal analysis of temporal variability in plant parameters has recently been carried out by Eghball & Power (1995) and Eghball et al (1995). The results of these two studies give an indication as to whether crop yields are dominated by long-term or short-term variation. Semivariogram models were used for the fractal analyses and the semivariance was calculated for a range of crops for different year intervals (*h*).

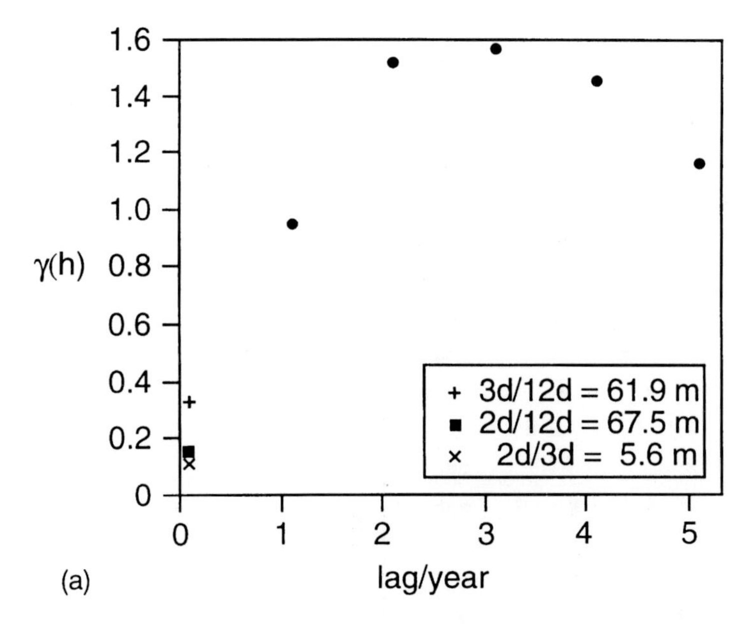

FIG. 2. Analyses of Rothamsted Park Grass and Broadbalk continuous wheat experiments. (a) Temporal variogram for Park Grass plots 2d, 3d and 12d. Spatial variances are also shown.

The fractal dimension (D) was calculated from the slope of the regression line in a plot of log semivariance versus log h, where $D = (4 - \text{slope})/2$. From equation (2) above the slope is equal to θ. As for spatial variation, a value of D close to 1 suggested that long-term variation dominated whereas a value approaching 2 suggested the dominance of short-term variation.

The advantage of using a fractal analysis over regression analyses to compare temporal variability between crops is that the results of regression analyses cannot be compared because the slope depends on yield levels and they would need to be standardized to allow comparison (Eghball & Power 1995). Fractal analysis is scale independent and the values of D depend on variability rather than yield, so values of D may be compared. D is a useful device for comparing the magnitude of spatial and temporal variation.

Eghball & Power (1995) analysed the average yields of barley, maize, oat, peanut, rice, rye, sorghum, soybean, wheat and cotton fibre in the USA over the 61 years from 1930 to 1990. Improvements in plant breeding and increased fertilizer, pesticide and herbicide use contributed to a strong increase in yields during this period. An analysis of covariance showed that there were significant differences between the slopes of the regression lines (calculated as above) for the 10 crops. The values of D ranged from 1.20 for rice to 1.47 for oats. Rice therefore had the least short-term variation whereas in oats and soybeans it was more pronounced, suggesting that these two crops may be particularly sensitive to annual variation in some growth factors.

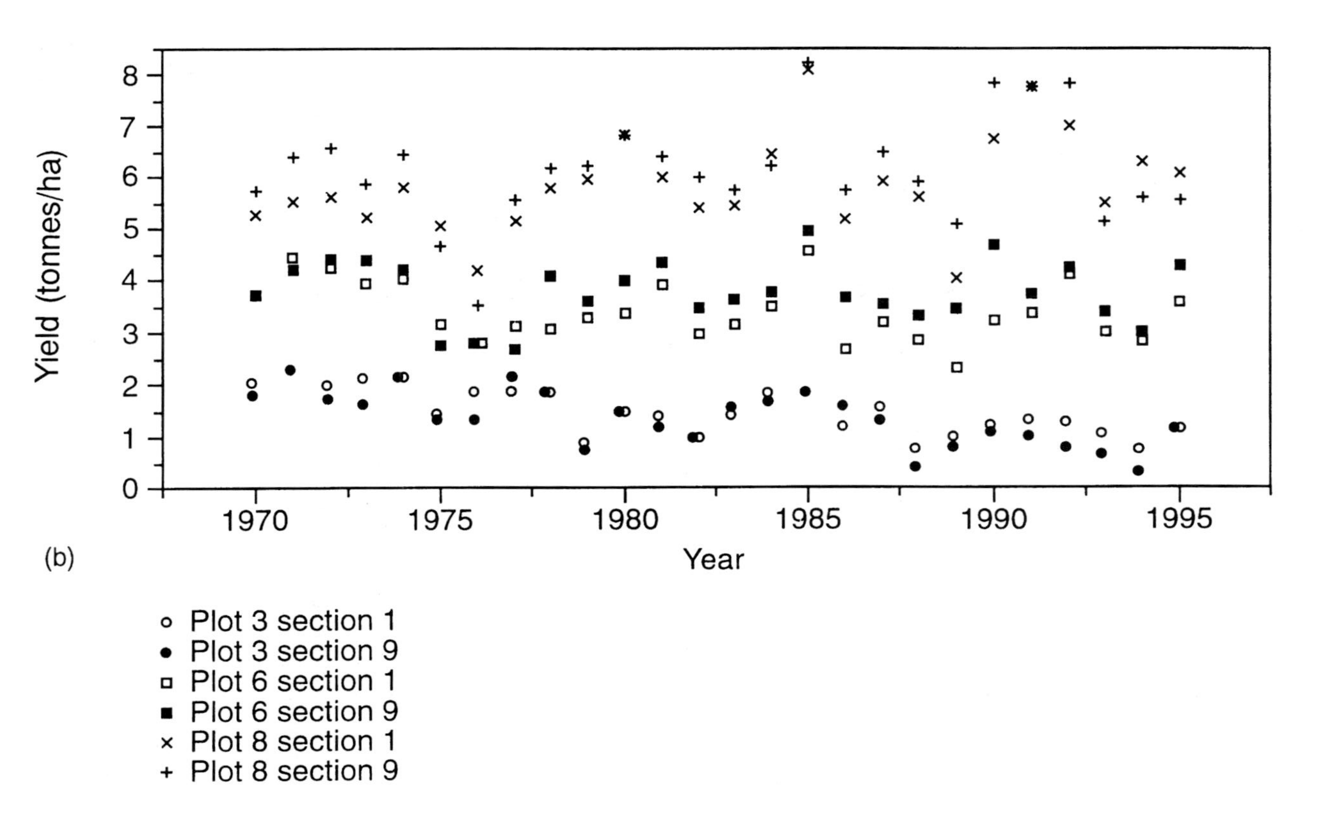

FIG. 2. Analyses of Rothamsted Park Grass and Broadbalk continuous wheat experiments. (b) Variation in wheat yield on plots 3, 6 and 8 of Broadbalk sections 1 and 9 from 1970 to 1995 after removal of varietal means.

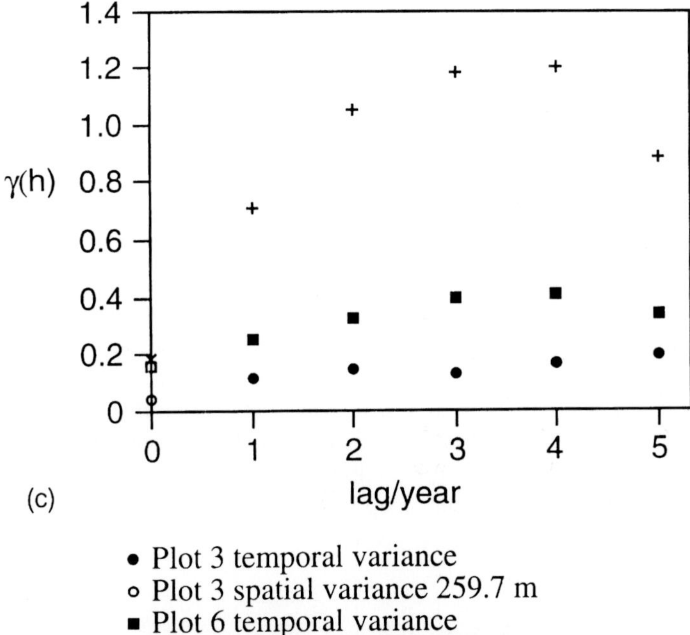

FIG. 2. Analyses of Rothamsted Park Grass and Broadbalk continuous wheat experiments. (c) Temporal variogram for three treatments on Broadbalk: plot 3 on sections 1 and 9 (unfertilized), plot 6 on sections 1 and 9 (N1, P, K) and plot 8 on sections 1 and 9 (N3, P, K).

Grain yield data (1953–1993) for maize under different management regimes were used in the temporal yield variability study by Eghball et al (1995). Plots were divided into manured and non-manured sections and to either section different fertilizer treatments (0, 45, 90, 135, 180 kg N/ha and 135 kg N + 80 kg P/ha) were applied. Values of D_v for the manured and non-manured plots were 1.971 and 1.981, respectively, and there was no significant difference between the two. This was also the case for the different fertilizer treatments with D values ranging from 1.958 to 1.996. These results indicate that short-term variation dominates. The authors conclude that for this study location (western Nebraska) management practices cannot override the strong influence of variable environmental conditions. In contrast, the study by Eghball & Power (1995) displayed long-term variability dominance which may have been influenced by the use of yield averages for the USA. Intuitively, we would expect greater fluctuation in yield data for smaller case studies.

In summary, high D values may indicate that environmental factors rather than management practices affect the year-to-year variability of crop yields. However, where long-term variability dominates it may be possible to predict crop yields over time and model the temporal plant growth (Eghball et al 1995). Values of D also indicate the risk involved in growing a particular crop in a particular location. Increased yield variability in the short-term (higher D) indicates greater risk in crop production.

Space–time models

The development of space–time models remains in its infancy, although several models have been suggested (e.g. Stein 1986, Posa 1993). Buxton & Pate (1994) have used a joint temporal/spatial variogram in a three-dimensional kriging process to estimate pollutant concentrations in time and space, the validity of their method being categorically confirmed by Dimitrakopoulos & Luo (1994). Heuvelink et al (1996) applied a more flexible model to the prediction of soil moisture under a pine forest.

Taking what is perhaps a simplistic approach, we can write general models as follows.

Stationary model:

$$\psi(x,y,t) = m_{x,y,t} + r(x,y,t) \tag{3}$$

where ψ is the yield, x, y and t are coordinates, m is a mean or trend, r is a residual and the subscripts may be taken to mean 'does not depend on the appropriate coordinates' and ( ) is taken to mean 'is dependent on the appropriate or associated coordinates'. So in the case above, yield is dependent on all coordinates but has a fixed mean. Several other models can also be written (Equations 4 & 5).

Non-stationary model:

$$\psi(x,y,t) = m(x,y,t) + r(x,y,t) \tag{4}$$

Intermediate models (e.g.):

$$\psi(x,y,t) = m_{x,y} + m(t) + r(x,y,t) \tag{5a}$$

$$\psi(x,y,t) = m(x,y,t) + r_{x,y}(t). \tag{5b}$$

Suitable models for the trend $m(x,y,t)$ would be generalized linear models e.g. polynomial trend surfaces (McCullagh & Nelder 1989), generalized additive models e.g. smoothing splines (Hastie & Tibshirani 1990), regression-tree models (Clark & Pregibon 1992) or a deterministic function describing a physical process. The usual model for residuals, $r(x,y,t)$, assumes the condition stated in Equation (6).

$$E[r(x,y,t)] = 0 \tag{6}$$

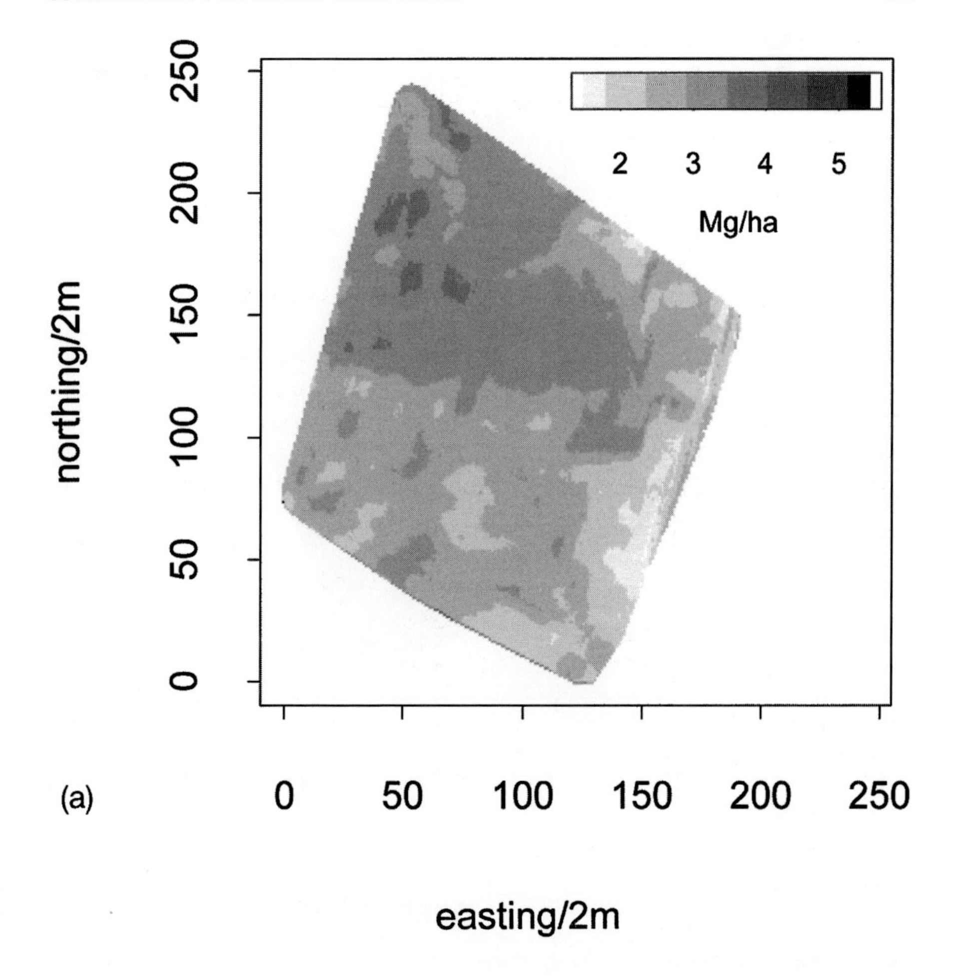

FIG. 3. Wheat yield from 'Horse' field, 'Marinya', Moree, NSW, Australia. (a) Yield map for December 1995.

Description of the residual is commonly performed by a covariance or semivariance function. Assuming spatial isotropy, one positive definite model for the semivariance (Christakos 1992) is shown as Equation (7).

$$\gamma(\lambda,\tau) = C_0 + C\left(1 - \exp\left(-\sqrt{\left(\frac{\lambda^2}{a^2} + \frac{\tau^2}{b^2}\right)}\right)\right) \tag{7}$$

Where $\gamma(\lambda,\tau)$ is semivariance as a function of the spatial lag (λ) and the temporal lag (τ), C_0 = nugget semivariance, C = sill semivariance minus nugget semivariance and a and b are selected to the range of influence in space and time respectively.

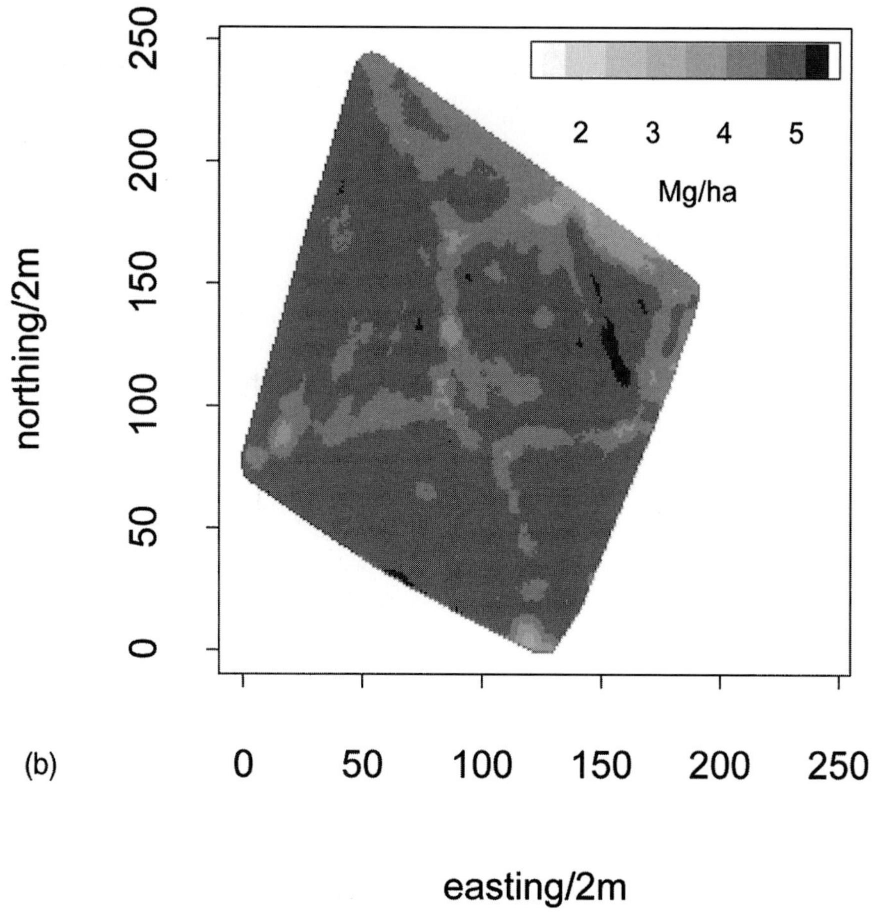

FIG. 3. Wheat yield from 'Horse' field, 'Marinya', Moree, NSW, Australia. (b) Yield map for December 1996.

Spatiotemporal examples

Rothamsted classical experiments

Data from the Rothamsted classical experiments were kindly provided by Rothamsted Experimental Station.

Park grass experiment. Dry matter data from three plots under identical management during the period 1965–1996 were analysed for variance over distance and time. Johnston (1994) provides a full description of the experiment. Analyses were done on total dry matter for each year, combining the dry matter yield from individual

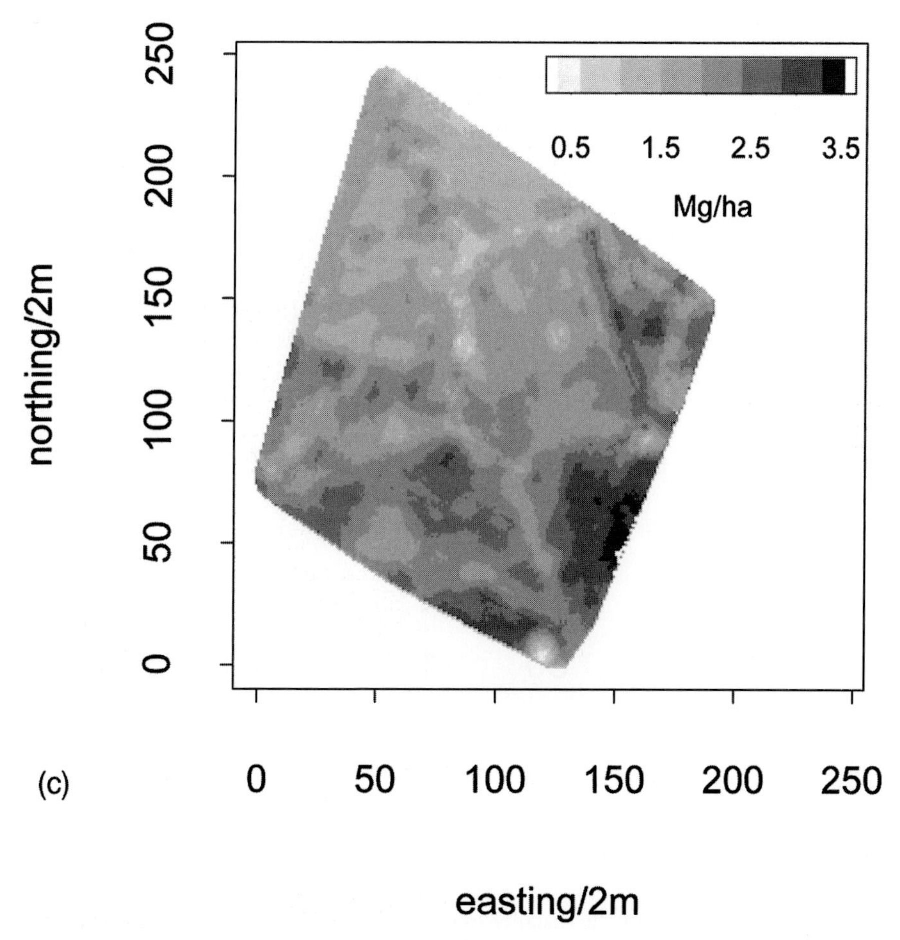

FIG. 3. Wheat yield from 'Horse' field, 'Marinya', Moree, NSW, Australia. (c) Map of yield differences between 1996 and 1995.

cuts. The average variances for a series of time lags and for the distance between plots were calculated. Figure 2a shows the average temporal variance for the three plots and also indicates the effect of spatial variance. Clearly, the temporal variance which peaks at 1.6 for a time separation of 3 years is much larger than the spatial variance (max = 0.35).

Broadbalk experiment. The spatial and temporal variability in wheat grain yield was analysed from continuous wheat treatments (sections 1 and 9) in plots 3, 6 and 8 for the period 1970–1995 (Fig. 2b). Each of these plots was under a different fertilizer treatment, as described by Johnston (1994). Average variances were calculated for a

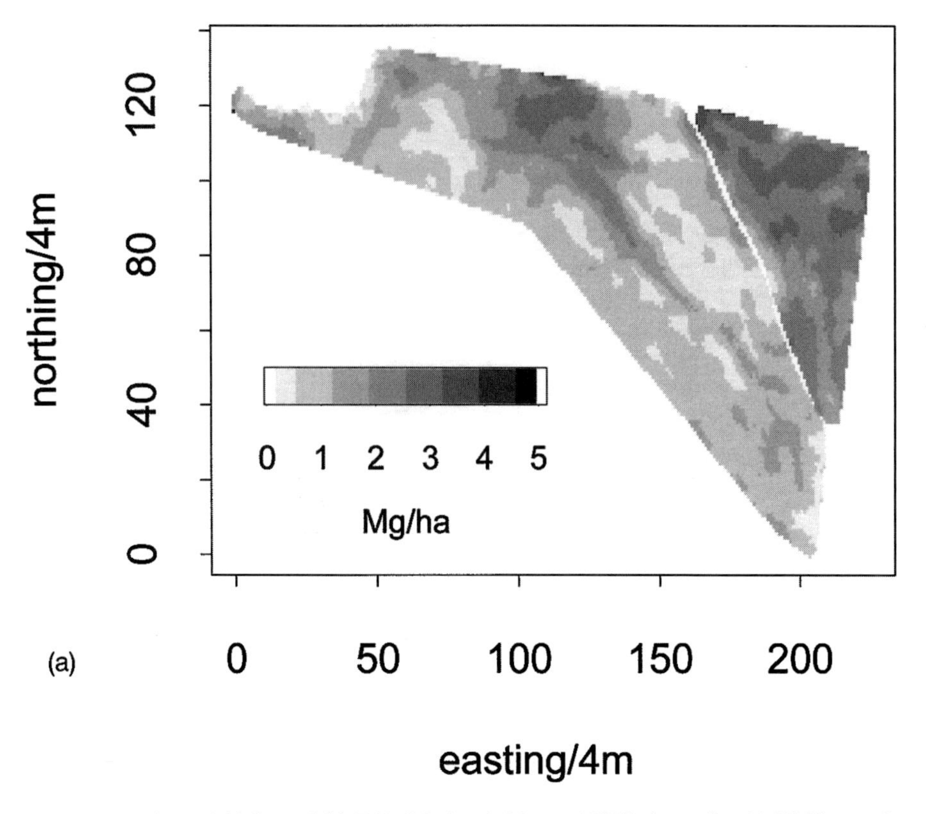

FIG. 4. Wheat yield from field 'B2', 'Marinya', Moree, NSW, Australia. (a) Yield map for December 1995.

series of time lags and for the distance between sections (259.7 m) for each treatment (Fig. 2c). It is evident from these figures that: (1) the treatment effect is large; (2) section 9 generally yielded more highly than section 1 across all treatments, suggesting an interaction with one or more yield-determining factors; and (3) within treatments the annual (temporal) variation is larger than the plot-to-plot (spatial) variation.

Marinya wheat yields

Wheat grain yields on two fields in NSW, Australia, were monitored using a combine mounted real-time grain yield sensor and Differential Global Positioning System receiver (DGPS) during the harvests of 1995 and 1996. Figures 3 & 4 separately display data for the two fields , 'Horse' and 'B2', and in each case show: (a) 1995 yield map; (b) 1996 yield map; and (c) difference in yield between 1996 and 1995. For mapping purposes a circular moving average with 10 m radius has been used to interpolate the data onto a grid of 2 m for 'Horse' and 4 m for 'B2'.

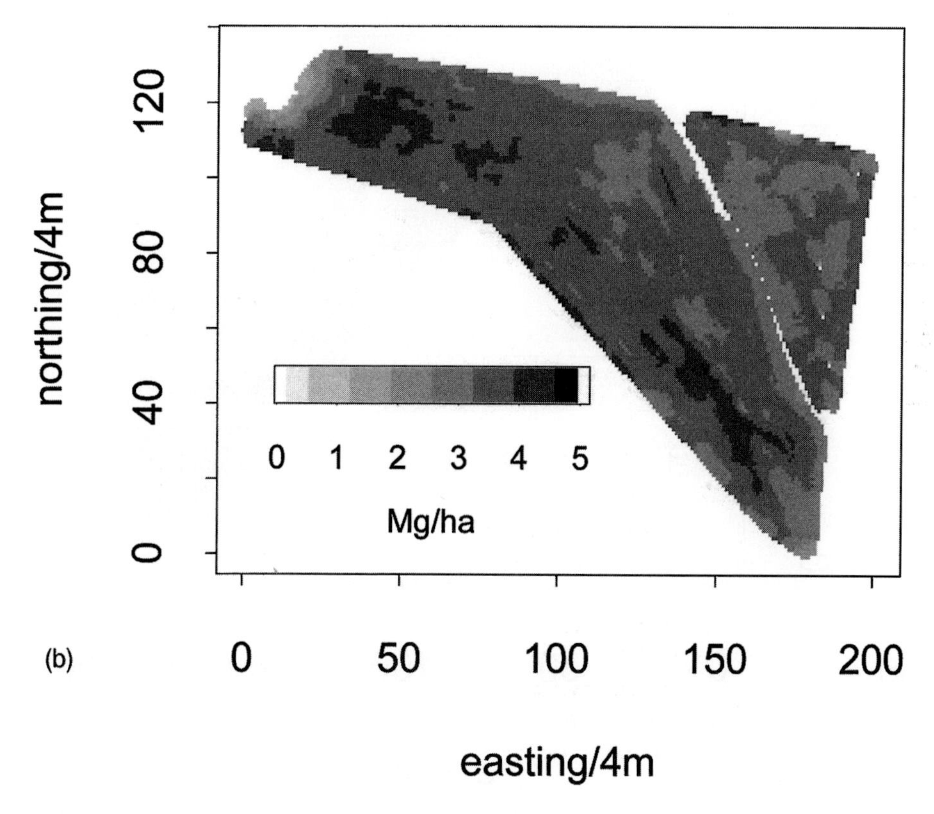

FIG. 4. Wheat yield from field 'B2', 'Marinya', Moree, NSW, Australia. (b) Yield map for December 1996.

The 1995 yield (Figs 3a & 4a) was significantly affected by late frost and harvest rainfall damage, whereas the 1996 season (Figs 3b & 4b) was completed under almost ideal growing conditions. Figure 3c shows the spatial and temporal yield differences for 'Horse' field and highlights a proportionally greater increase for the southern half. Figure 4c displays the spatial and temporal yield differences for field 'B2'.

Table 2 documents the moments for the data in Figs 3 and 4. Once again the spatial variation in the fields appears smaller than the year to year (temporal) variation. On 'Horse' the 1996 yield is higher than 1995 at all points in the field. However, on 'B2' there are areas which have smaller yields in 1996 than 1995 and other areas which have significantly higher yields in 1996 than in 1995, suggesting there may be some merit in subzoning the field for management purposes.

The Jyndevad long-term barley experiment

Ersbøll (1994) examined the spatial and temporal variance in grain yield for a 14-year continuous spring barley experiment conducted at Jyndevad Experimental Station,

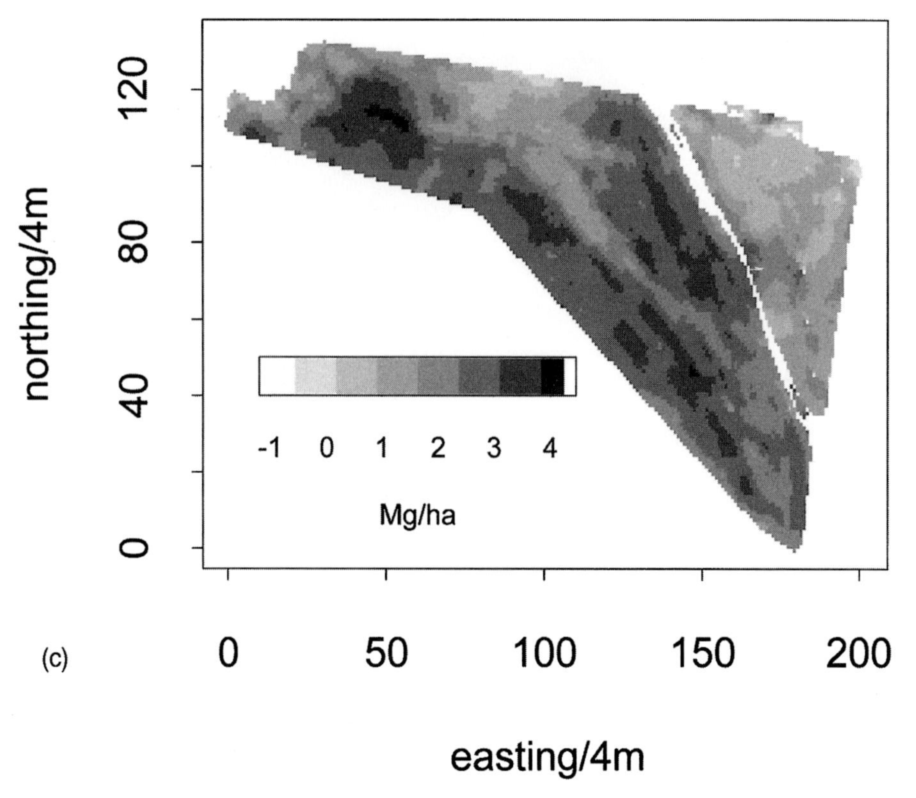

FIG. 4. Wheat yield from field 'B2', 'Marinya', Moree, NSW, Australia. (c) Map of yield differences between 1996 and 1995.

Denmark (mean yield 3.4 t/ha). After removal of tillage and fertilizer treatment effects, the residuals for each year were calculated and spatial semivariograms estimated. Deviations in estimates of the sill ($C_0 + C$) were found to occur between years, indicating significant temporal variation, while the range a remained relatively constant, suggesting a more uniform spatial variation over time.

In order to jointly estimate the contribution of space and time to yield variance we fitted a spatiotemporal semivariogram model (Equation 8; a modification of Equation 7) to the empirical spatiotemporal semivariogram.

$$\gamma(\lambda,\tau) = C_0 + C\left(1 - \exp\left(-\sqrt{\left(\frac{\lambda}{a} + \frac{\tau}{b}\right)}\right)\right) \tag{8}$$

The ranges of influence were estimated to be: spatial = 77.6 m; temporal = 1.3 years. The nugget effect was estimated to be 0.67 $(t/ha)^2$ and the sill at 1.52 $(t/ha)^2$. These results again display low annual correlation and reasonably substantial spatial

TABLE 2 Marinya wheat yields: descriptive statistics of two fields in NSW, Australia, monitored using real-time yield sensors

(a) Horse field: wheat yields (tonnes/ha)

Year	*Mean*	*Variance*	*Standard deviation*	*CV*
1995	2.73	0.195	0.441	0.162
1996	4.66	0.081	0.285	0.061
1995–1996	−1.93	0.252	0.502	
Temporal variance		1.99		

(b) B2: wheat yields (tonnes/ha)

Year	*Mean*	*Variance*	*Standard deviation*	*CV*
1995	1.32	0.558	0.747	0.564
1996	3.43	0.156	0.394	0.114
1995–1996	−2.14	0.747	0.864	
Temporal variance		2.67		

correlation suggesting that temporal variation dominates the yield over the 14 year period.

Spatiotemporal modelling of volumetric soil water content

In the Eastern Australian cropping belt, soil moisture is the most important single factor controlling crop yield. From 1993–1995 we established a field experiment to continuously monitor and describe soil water content at a small scale during the growth of a wheat crop and to use the soil moisture data and models to help explain variation in crop yield at a fine-scale. The study also aimed to examine the correlation between scale in soil variation and scale in yield variation.

The experiment was conducted on a 36 m × 36 m site subdivided into 144 3 m × 3 m plots each with a centrally located multiplexed time-domain reflectometry (TDR) waveguide (20 cm long) inserted horizontally at a depth of 25 cm. Ten additional waveguides (5 pairs at 30 cm separation) were used to examine shorter-range variation (Fig. 5a). Experiments were conducted on a pelustert (~60% clay).

Table 3 presents a description of the data for 1993 and 1995 (insufficient rainfall precluded sowing in 1994). To define the trend component of the data a number of models described earlier were applied and tested for goodness of fit (Table 4). The most successful of these, the regression-tree model, is a hierarchical model that

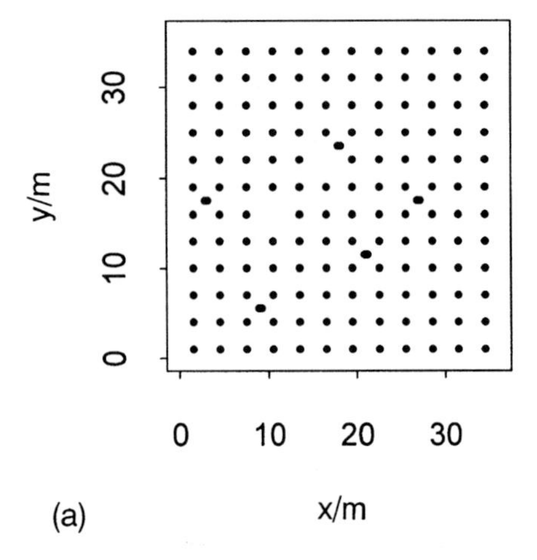

FIG. 5. (a) Location of time-domain reflectometry waveguides.

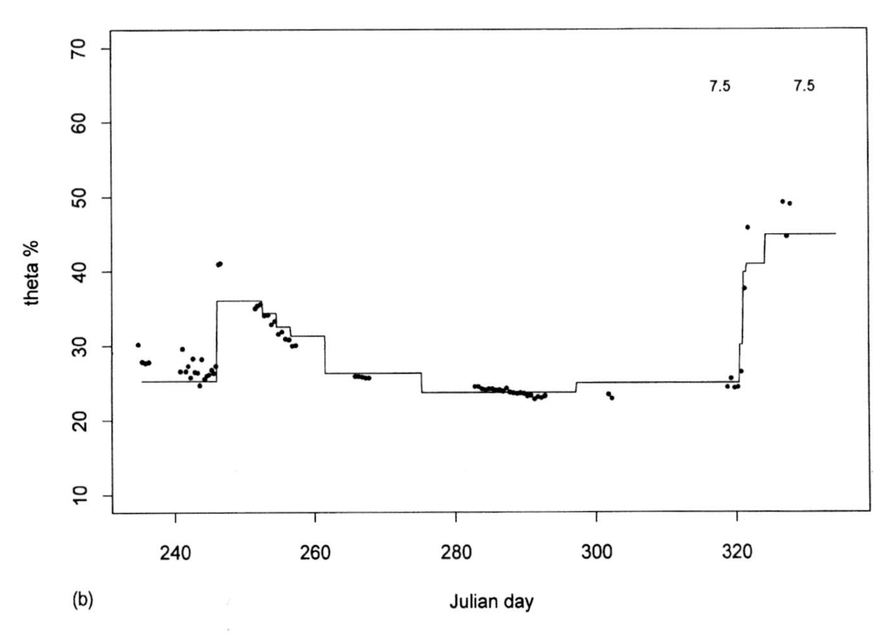

FIG. 5. (b) Variation in volumetric soil water content (θ) in 1993 at site x = 7.5, y = 7.5 (depth = 25 cm). Solid discs are observations, the line is a fitted regression-tree model.

determines the relationship between variables by devising a set of regression rules for prediction using recursive partitioning. Figure 5b presents the model fit for one measurement point over the 1993 growing season and Fig. 5c shows the model fit for the whole site at one point in time (September 4, 1993).

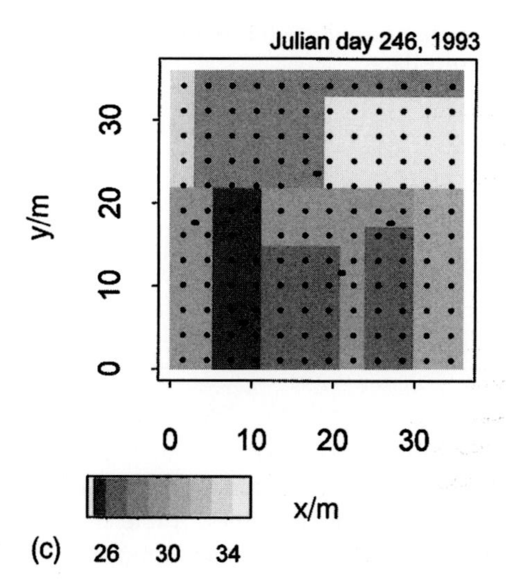

FIG. 5. (c) Fit of regression-tree model on Julian day 246 in 1993.

TABLE 3 Descriptive statistics for soil moisture monitoring experiment

Moment	*θ% 1993*	*θ% 1995*
minimum	15.0	13.1
mean	33.7	30.0
maximum	66.2	59.6
s.d.	6.9	6.1
variance	47.6	37.1
duration	99 days	105 days

TABLE 4 Goodness of fit for trend models fitted to soil moisture data

	% Variance explained	
Model	*1993*	*1995*
Polynomial trend surface	24.4	1.1
Generalized additive model	32.0	37.0
Regression tree	62.9	66.9

TABLE 5 Parameter estimates for space–time variogram model of residuals for moisture data

Parameter	*1993*	*1995*
C_0	15	8
C	3	4
a	17 m	42 m
b	6.5 days	8 days
a/*b*	2.6 m/day	5.2 m/day

A spatiotemporal variogram model (Equation 7) fitted to the regression-tree residuals completed the modelling procedure and allowed spatial prediction of soil moisture across the site at unsampled times. Table 5 catalogues the variogram parameters for the two years, showing quite short temporal correlation.

It was concluded that large variation in both space and time made predictions problematic, and although it was a reasonable first attempt, the regression-tree/kriging model did not describe the data adequately. A more physical type of model is required that may include rainfall events and other external conditions with intermittent exponential temporal decay. An approach to such a model has been described by Or & Hanks (1992).

Discussion and conclusion

There's an old saying in farming that it is not cleanliness but timeliness that is next to godliness. What does this say about 'spaceliness'? Certainly, the data presented above suggest that annual temporal variation is much larger than yield variation over single fields. But there was also some evidence of (non-linear) interaction between parts of fields and the environment to produce different effects in different years. If all yields in the field simply increased or decreased from year to year then the advantage of spaceliness, spatial intervention or differential treatment would be negligible.

The experimental modelling to date has shown that the level of uncertainty in space–time crop yield data remains difficult to assess reliably. However, we have not discussed the soil/environment/crop interactions that underlie yield response variability. To fully encapsulate this variation it will be important to know the spatial and temporal scale of yield determining factors such as insects, soil-borne pathogens, weather, nutrients, water and weeds. Interactions with these production variables will be crop-specific to some degree and not all will be manageable (at least economically).

Practitioners of precision agriculture are now attempting to optimize—both economically and environmentally—the treatment of such spatial and temporal variation in cropping-system components. At present the scale at which these individual crop growth parameters vary within the field and their true impact on yield variability is undetermined. This governs the uncertainty in differential treatment decisions.

In conclusion, we pose the following null hypothesis that remains to be seriously researched for the successful application of precision agriculture: 'given the large temporal variation evident in crop yield relative to the scale of a single field, then the optimal risk aversion strategy is uniform management'. Certainly we need to investigate under what particular circumstances this will not be the case.

Acknowledgements

We thank IACR Rothamsted for permission to use data from the Park Grass and Broadbalk Classical Experiments, and in particular A. E. Johnston and D. Yeomans for extracting the data and sharing their knowledge of the experiments. We also thank Craig and Judy Boydell for access to, and help with, their wheat crop and Tom Bishop for aid in graphical presentations.

References

Berry MV, Lewis ZV 1980 On the Weierstrauss–Mandelbrot fractal function. Proc R Soc Lond Ser A 370:459–484

Bresler E, Dasberg S, Russo D, Dagan G 1981 Spatial variability of crop yield as a stochastic soil process. Soil Sci Soc Am J 45:600–605

Bresler E, Dagan G, Russo D 1982 Statistical analysis of crop yield under controlled line-source irrigation. Soil Sci Soc Am J 46:841–847

Burrough PA 1981 Fractal dimensions of landscapes and other environmental data. Nature 294:240–242

Buxton BE, Pate AD 1994 Joint temporal–spatial modelling of concentrations of hazardous pollutants in urban air. In: Dimitrakopoulos R (ed) Geostatistics for the next century. Kluwer Academic, The Netherlands, p 75–87

Carr PM, Carlson GR, Jacobsen JS, Nielsen GA, Skogley EO 1991 Farming soils, not fields: a strategy for increasing fertiliser profitability. J Prod Agric 4:57–61

Christakos G 1992 Random field models in earth sciences. Academic Press, Troy, MO

Clark LA, Pregibon D 1992 Tree-based models. In: Chambers JM, Hastie TJ (eds) Statistical models in S. Chapman & Hall, New York, p 377–420

Dimitrakopoulos R, Luo X 1994 Spatiotemporal modelling: covariances and ordinary kriging systems. In: Dimitrakopoulos R (ed) Geostatistics for the next century. Kluwer Academic, The Netherlands, p 88–93

Eghball B, Power JF 1995 Fractal description of temporal yield variability of 10 crops in the United States. Agron J 87:152–156

Eghball B, Binford GD, Power JF, Baltensperger DD, Anderson FN 1995 Maize temporal yield variability under long-term manure and fertiliser application: fractal analysis. Soil Sci Soc Am J 59:1360–1364

Ersbøll AK 1994 On the spatial and temporal correlations in experimentation with agricultural applications. PhD thesis, Institute of Mathematical Modelling, The Technical University of Denmark, Lyngby, Denmark

Guitjens JC 1992 Interpreting spatial yield variability of irrigated spring wheat. Trans ASAE (Am Soc Agric Eng) 35:91–95

Hastie TJ, Tibshirani RJ 1990 Generalised additive models. Chapman & Hall, London

Heuvelink GBM, Musters P, Pebesma EJ 1996 Spatio-temporal kriging of soil water content. Proceedings of the International Geostatistics Conference, Wollongong. Kluwer Academic, The Netherlands, p 1020–1030

Hunsaker DJ 1992 Cotton yield variability under level basin irrigation. Trans ASAE (Am Soc Agric Eng) 35:1205–1211

Johnston AE 1994 The Rothamsted classical experiments. In: Leigh RA, Johnston AE (eds) Long-term experiments in agricultural and ecological sciences: proceedings of a conference to celebrate the 150th anniversary of Rothamsted Experimental Station, held at Rothamsted, 14–17 July 1993. CAB International, Wallingford, p 9–37

Journel AG, Huijbregts CJ 1978 Mining geostatistics. Academic Press, London

McCullagh P, Nelder JA 1989 Generalised linear models, 2nd edn. Chapman & Hall, London

Miller MP, Singer MJ, Nielsen DR 1988 Spatial variability of wheat yield and soil properties on complex hills. Soil Sci Soc Am J 52:1133–1141

Or D, Hanks RJ 1992 Spatial and temporal soil water estimation considering soil variability and evapotranspiration uncertainty. Water Resource Res 28:803–814

Posa D 1993 A simple description of spatial–temporal processes. Comp Stat Data Analysis 15:425–437

Smith HF 1938 An empirical law describing heterogeneity in the yields of agricultural crops. J Agric Sci 28:1–23

Stein M 1986 A simple model for spatial–temporal processes. Water Resource Res 22:2107–2110

Spatial sampling

Steven K. Thompson

Department of Statistics, Pennsylvania State University, 326 Classroom Building, University Park, PA 16802–2111, USA

Abstract. Sampling in a spatial setting is typically done for the purpose of estimating or predicting a population quantity such as the total of a variable in a study region, to predict a new value at an unobserved site, or to find regions of high values for further study. Spatial covariance and conditional variance patterns have led to the use of designs such as systematic and stratified arrangements to increase precision and cluster or multistage sampling to increase cost effectiveness. In addition, adaptive procedures can be used to take advantage of patterns not discovered prior to the survey. An adaptive sampling design is one in which the procedure for selecting sites or units to include in the sample can depend on values of the variable of interest observed during the survey. For example, in a survey of an unevenly distributed insect species, neighbouring sites may be added to the sample whenever high abundance of the insects is encountered. Examples of adaptive designs include adaptive cluster sampling and adaptive allocation. Issues of design optimality or effectiveness and choice of inference methods with different types of designs will be discussed.

1997 Precision agriculture: spatial and temporal variability of environmental quality. Wiley, Chichester (Ciba Foundation Symposium 210) p 161–172

Sampling to estimate a total, predict new values, or find high values

In a spatial setting, the objective of sampling may be (1) to estimate a total, such as the total abundance of an insect pest in a field, (2) to predict a new value, such as abundance of the insects at a site not yet observed, or (3) to find high values, either to locate hot spots of pest activity or simply to obtain a large enough number of individuals of the species for observation. To accommodate these objectives, the sampling problem has two parts: (1) selecting the sample of units or sites at which to make observations, and (2) making estimates, predictions, or other inferences about the population using the observed sample data. The reason for sampling is that most often it is impractical to observe the variables of interest over the entire study region. Estimating the population total from a sample or predicting the value at an unobserved site is made difficult by the fact that many of the variables of interest — such as insect abundance in a field — are extremely uneven spatially. For example, Fleischer et al (1996), in discussing problems of site-specific management of an insect population, note that not only is insect abundance highly variable within a field and among fields, but also that key traits such as genetic resistance to insecticides also exhibit high spatial variability within the population.

To overcome the inherent spatial variability one would like to use large sample sizes. However, in spatial sampling situations often the observations or measurements are difficult or expensive to obtain, as in determining abundances of organisms in the soil or of trace environmental contaminants. Thus, large sample sizes are often not the answer, and one must therefore look for the most efficient possible sampling designs. The search for efficient sampling strategies for spatial studies with high, unpredictable variability is the topic of this paper.

Systematic and stratified designs are often efficient

The sampling design is the procedure by which a sample s of units or sites is selected. A probability design designates, for each possible sample s, the probability of selecting that sample. For example, a simple random sampling design of size n gives equal probability to every possible collection of n sites. However, in the spatial setting with the characteristically patchy distributions of variables such as insects, simple random sampling is not a notably efficient design for either estimating a total or predicting a new value, because random samples of sites tend themselves to be very unevenly distributed in space.

What design will be most effective depends on the tendencies of the population. With many natural populations, nearby sites tend to have some similarity to each other, while sites distant from each other are more nearly independent (Matérn 1986). The spatial covariance function for such a population is thus positive (or non-negative) and decreasing with distance.

When the objective is to estimate a population total or mean over the study region, for which the most common form of estimator involves a linear combination of the sample data, the variance of the estimator is a sum including every possible covariance between pairs of sample sites. For the most precise estimate we would therefore like these covariances to be as small as possible, which for populations with non-negative, decreasing covariance functions as so frequently found in nature is accomplished by spacing the sample units as far apart as possible, on average, by using a systematic design (McBratney & Burgess 1981a,b, Yfantis et al 1987, McArthur 1987).

When the objective is to accurately map the variable over the entire study region, as for site-specific management of an insect pest species, one criterion is to seek to minimize the largest mean square prediction error among the predicted values. Since the maximum distance from any location to the nearest sample site tends also to be minimized by spacing sites apart, systematic designs tend also to be efficient for the mapping problem. The actual optimization problem is somewhat more complex when one considers the boundary of the study region and the possible additional objective of estimating the spatial covariance function or variogram from the sample data (Corsten & Stein 1994). On the other hand, a systematic design is very inefficient when the systematic grid coincides with some existent periodicity in the population, as for example when a systematic selection of sites for assessing insect abundances on plants in a field results in all sites coinciding with watering system outlets. One way to avoid periodicity issues and other possible biases while still tending to space sample sites apart is to use stratified random sampling, in which the field is partitioned into many spatial strata and a random sample of two or more sites is selected in each of the strata.

With stratified sampling, further efficiency is gained by allocating relatively more sampling effort to strata in which the variable of interest is more variable. With many natural populations, variability tends to be high where abundance is high, so that efficiency is gained by allocating more sampling effort where abundance is high and less effort where abundance is low.

The optimal design is often an adaptive one

For populations such as insects whose spatial distributions are highly clustered and uneven, but unpredictably so, one has insufficient information prior to the survey to allocate more sampling effort to regions of high abundance or variability. Situations such as this have motivated the development of adaptive designs, in which the procedure for selecting the sample responds to values of the variable of interest observed during the survey. For example, in a survey of a patchily distributed insect species, an initial systematic sample might be selected, with nearby sites added to the sample whenever high abundance is observed.

From a theoretical point of view it can be shown that the optimal sampling strategy is often an adaptive one (Zacks 1969, Thompson 1988, Thompson & Seber 1996). Suppose one has a stochastic model for the population and wishes to predict a quantity T, such as the region total or the value at an unobserved site, using a predictor $\widehat{T}$ based on a sample of n sites. After selecting an initial sample s_1 of n_1 sites and observing the values, denoted collectively $\mathbf{y}_{s_1}$, of the variable of interest at those sites, the optimal selection of the additional $(n-n_1)$ sites is the sample s_2 which minimizes the conditional mean square error

$$g_{s_2}(s_1,\mathbf{y}_{s_1}) = \mathrm{E}[(\widehat{T} - T)^2 | s_1,\mathbf{y}_{s_1}]$$

In words, the optimal choice is to choose the remaining sample to minimize the conditional mean square error of the predictor given the observations made so far. Note that the predictor $\widehat{T}$ is a function of the full sample and values observed, that is, $\widehat{T} = \widehat{T}(s_1, s_2, \mathbf{y}_{s_1}, \mathbf{y}_{s_2})$, which is why the conditional expectation may differ for different choices of s_2. By minimizing the conditional mean square error, the unconditional mean square error $E(\widehat{T} - T^2)$ is minimized. The above reasoning using two phases can be generalized to make the optimal choice sequentially for each of the n units.

If a single s_2 minimizes the function g for every value of $\mathbf{y}_{s_1}$, then the optimal strategy is a conventional one, in which the entire sample can be selected prior to the survey. If, on the other hand, different choices of s_2 minimize g for different values of $\mathbf{y}_{s_1}$, the optimal strategy is an adaptive one. Examples of spatial models in which the optimal strategy is conventional include models such as Gaussian white noise and the Poisson process in which values at separate sites are independent, and Gaussian processes of known covariance structure in which conditional covariances are independent of observed values (Zacks 1969, Thompson & Seber 1996). With other spatial models, such as log-Gaussian processes and Poisson cluster (Cox) processes, the optimal strategy is adaptive. Even with Gaussian processes an adaptive procedure may have an advantage when covariance structure is not known prior to the survey.

A problem with theoretically optimal strategies is that they may be impractical to implement in practice (Solomon & Zacks 1970).

The optimal strategy may require an unrealistic amount of prior information about the population, may be computationally complex, and may require measurement and analysis of observations faster than is possible.

Adaptive strategies that are simple, if not optimal

Two classes of adaptive strategies that are simple to implement, have easy-to-compute estimates, and are relatively free of assumptions about the population under study are adaptive cluster sampling and adaptive allocation. While not theoretically optimal, these procedures can produce substantial gains in efficiency or precision compared to conventional designs of equivalent sample size in some spatial sampling situations.

Adaptive cluster sampling

In adaptive cluster sampling, an initial sample is selected using a conventional sampling design. Whenever the value of the variable of interest satisfies a specified condition, neighbouring units are added to the sample. For example, in sampling an insect species in a field, whenever abundance above a specified threshold level is observed on a plant in the sample, neighbouring plants are added to the sample and observed. If in turn any of the neighbouring plants has insects above the threshold level, additional neighbours are added, and so on.

The initial sample may be selected by any conventional sampling design, such as random sampling with or without replacement (Thompson 1990), conventional cluster sampling or systematic sampling (Thompson 1991a), stratified random sampling (Thompson 1991b), plotless unequal probability selection as used in forestry surveys (Roesch 1992), unequal probability sampling using an auxiliary variable (Smith et al 1995), and two-stage sampling (M. Salehi & G. A. F. Seber, unpublished observations 1996). Adaptive cluster sampling designs with multivariate variables of interest (Thompson 1993), based on order statistics (Thompson 1996), and with imperfect detectability (Thompson & Seber 1994) have also been described.

A conventional estimator such as a sample mean or expansion estimator, which is unbiased with a conventional design such as simple random sampling, will not be

TABLE 1 Comparative efficiencies of adaptive strategies to comparable conventional strategies

Population	*Adaptive*	*Conventional*	*Relative efficiency*	*Source*
Shrimp	aa	strat ran	1.2–1.3	Thompson et al 1992
Roughy	aa	strat ran	1.8	Francis 1984
Mackerel	aa	strat ran	1.27	Francis 1984
Trees	acs	uneq prob	1.0–2.1	Roesch 1993
Waterfowl	acs	srs, ppx	0.6–7.5	Smith et al 1995
Point process	acs (srs)	srs	0.5–15.4	Thompson 1990
Point process	acs (cluster)	cluster	1.0–4.9	Thompson 1991a
Point process	acs (sys)	sys	1.5–∞	Thompson 1991a
Point process	acs (strat)	strat ran	1.1–50.1	Thompson 1991b

Relative efficiency is given by the variance of the estimator with the conventional strategy divided by the variance of the estimator with the adaptive strategy. A value greater than one indicates that the adaptive strategy was more efficient. For each population the adaptive and conventional strategies compared are listed, followed by the relative efficiency and the source of the comparison. Adaptive strategies are adaptive allocation (aa) and adaptive cluster sampling (acs) with initial design in parentheses. Conventional designs are stratified random (strat ran), unequal probability (uneq prob), simple random sampling (srs), sampling with probabilities proportional to an auxiliary variable (ppx), cluster sampling and systematic sampling (sys).

unbiased with adaptive cluster sampling. However, simple unbiased estimators are available for use with adaptive cluster sampling (see Thompson 1992, Thompson & Seber 1996).

For rare, clustered populations, adaptive cluster sampling can produce substantial gains in efficiency compared with conventional designs of equivalent sample size (see Table 1).

Adaptive allocation

In stratified sampling, greatest precision is obtained for a given total sample size using optimal allocation formulas (cf. Cochran 1977, Thompson 1992). The formulas allocate relatively more of the sampling effort to strata which are more variable. When stratum standard deviations are not known prior to the survey, however, a natural alternative is to take an initial stratified sample and then allocate the rest of the sampling effort adaptively based on the initial sample standard deviations or other observed statistics.

Interestingly, when the allocation is made adaptively, the conventional stratified sampling estimators, which are unbiased with a conventional design, are no longer unbiased, though the biases may tend to be small (Francis 1984). A number of procedures are available for producing unbiased estimates with adaptive allocation, including basing the allocation on observations in adjacent strata (Thompson &

Ramsey 1983, Thompson et al 1992), applying the Rao-Blackwell method to the initial stratified estimator, or using 'fixed weight' estimators (Thompson & Seber 1996). Adaptive allocation designs can give reductions in mean square error or variance compared to fixed-allocation designs in some situations (see Table 1).

Is the design relevant in making estimates?

The unbiased estimators described above for adaptive cluster sampling and adaptive allocation are *design unbiased*, meaning that the average value of the estimator over all possible samples that might be selected equals exactly the actual population value. Thus the unbiasedness does not depend on any assumptions about the population itself. Regardless of whether the population is clustered, evenly spaced, or randomly distributed in the study region, the estimator is unbiased because of the design used in selecting the sample.

In contrast, the 'best linear unbiased' or 'kriging' estimator or predictor based on the variogram or covariance function with an assumed spatial model is said to be *model unbiased* because the unbiasedness is based on the assumed population model rather than on the sampling design. In fact, under the assumed model, the predictor is said to be unbiased for the specific sample selected, that is $E(\widehat{T}\,|s) = E(T)$. The practical advantage of such a predictor in the spatial setting is that when the sample is unevenly spaced in the study region, the predictor weights the observations to make the best prediction of the unobserved values given the particular sample. The kriging predictor takes the locations of the sample sites into account but ignores how the sample was selected.

An important question, however, is whether the properties of a model-based estimator or predictor are in fact influenced by how the sample was selected, or, put another way, whether even with the assumed model the design should be taken into account at the inference stage. Various aspects of this question have been examined by Basu (1969), Krieger & Pfeffermann (1992), Little (1982), Rubin (1976), Scott (1977), Sugden & Smith (1984) and Thompson & Seber (1996).

To answer that question it is useful to factor the likelihood function for the sampling outcome into design and model parts.

The design is designated $p(s|\mathbf{y}_s, \mathbf{y}_{\bar{s}}; \gamma)$, giving the probability of selecting the sample s. Potentially the selection probability may depend on values $\mathbf{y}_s$ of the variable of interest at sites in the sample, on values $\mathbf{y}_{\bar{s}}$ of the variable of interest at sites not in the sample, or on unknown parameter values γ. With a conventional design, such as simple random sampling or systematic sampling, the design is simply $p(s)$, since selection does not depend on any values of the variable of interest or any unknown parameter values. With an adaptive design such as adaptive cluster sampling or adaptive allocation the design is of the form $p(s|\mathbf{y}_s)$, depending on values of the variable of interest but only in the sample. Non-standard sampling procedures, such as happens when researchers look over the entire study region and then choose as the sample those units which look most 'representative' or even those units with the highest values — as when insects are collected only from the sites with most abundance—are of the form $p(s|\mathbf{y}_s,\mathbf{y}_{\bar{s}})$, depending on values both inside and outside of the sample.

Assuming the study region consists of N spatial units which potentially could be sampled, a population model is given by the joint distribution (probability density function) $f(\mathbf{y};\theta)$, which may depend on one or more unknown parameters θ. The relevant likelihood function can be described as the joint probability (density) that the realized population values are $\mathbf{y}$ and the sample s is selected. This is given by

$$L(\mathbf{y}_{\bar{s}},\gamma,\theta;\, s,\mathbf{y}_s) = p(s|\mathbf{y}_s,\mathbf{y}_{\bar{s}};\, \gamma) f(\mathbf{y}_s,\mathbf{y}_{\bar{s}};\, \theta)$$

That is,

$$\text{Likelihood} = (\text{Design}) \times (\text{Model})$$

The design expectation, which determines whether an estimator $\widehat{T}$ is design unbiased, is

$$\mathrm{E}(\widehat{T}|\mathbf{y}) = \sum_{\text{all } s} \widehat{T}_s p(s|\mathbf{y}_s,\mathbf{y}_{\bar{s}};\, \gamma)$$

where $\widehat{T}_s$ designates the value of the estimate for the given sample s. The estimator $\widehat{T}$ is design unbiased if $\mathrm{E}(\widehat{T}|\mathbf{y}) = T$, no matter what the realization $\mathbf{y}$.

The model expectation of $\widehat{T}$ is $\mathrm{E}(\widehat{T}|s)$, which can be shown to be

$$\mathrm{E}(\widehat{T}\,|\,s) = \frac{\int \widehat{T}(s,\mathbf{y}_s) p(s\,|\,\mathbf{y}_s,\mathbf{y}_{\bar{s}};\, \gamma) f(\mathbf{y}_s,\mathbf{y}_{\bar{s}};\, \theta)\, d\mathbf{y}}{\int p(s\,|\,\mathbf{y}_s,\mathbf{y}_{\bar{s}};\, \gamma) f(\mathbf{y}_s,\mathbf{y}_{\bar{s}};\, \theta)\, d\mathbf{y}}$$

Now, if the design is a conventional one, so that $p(s\,|\,\mathbf{y}_s,\mathbf{y}_{\bar{s}};\gamma) = p(s)$ then the design comes out of the integrals and cancels out, the denominator integrates to one, and so the expectation depends only on the assumed model, not on the design. With any adaptive or non-standard design, however, the model expectation depends on the design.

Thus, with a selection procedure that depends on values of the variable of interest, the model-based expectation of an estimator depends not only on the assumed model, but on the design as well. Therefore, with an adaptive or non-standard site-selection procedure, standard model-based estimators, such as the expanded sample mean or the kriging predictor which are model unbiased with a conventional design, are not unbiased even if the assumed model is true. A design-unbiased estimator, on the other hand, will be unconditionally unbiased under any population model.

Likelihood-based inference, such as Bayes and maximum-likelihood estimation and prediction, on the other hand is not affected by an adaptive design depending only on $\mathbf{y}_s$ but is affected by a non-standard selection procedure depending on $\mathbf{y}_{\bar{s}}$ or on an unknown parameter γ, if γ is in any way related to the model parameter θ. These results follow immediately from the factorization of the likelihood into design and model components. Since kriging predictors have a maximum likelihood interpretation with an assumed Gaussian model, the maximum likelihood status of these predictors is therefore unaffected by an adaptive design.

References

Basu D 1969 Role of the sufficiency and likelihood principles in sample survey theory. Sankhya 31:441–454

Cochran WG 1977 Sampling techniques, 3rd edn. Wiley, New York

Corsten LCA, Stein A 1994 Nested sampling for estimating spatial semivariograms compared to other designs. Appl Stoch Models Data Analysis 10:103–122

Fleischer SJ, Weisz R, Smilowitz A, Midgarden D 1996 Spatial variation in insect populations and site-specific management. In: Pierce FJ, Sadler EJ (eds) The state of site-specific management for agriculture. ASA-CSSA-SSSA, Madison, WI

Francis RICC 1984 An adaptive strategy for stratified random trawl surveys. N Z J Mar Freshwater Res 18:59–71

Krieger AM, Pfeffermann D 1992 Maximum likelihood estimation from complex sample surveys. Surv Method 18:225–239

Little RJA 1982 Models for nonresponse in sample surveys. J Am Stat Assoc 77:237–250

Matérn B 1986 Spatial variation, 2nd edn. Springer-Verlag, Berlin

McArthur RD 1987 An evaluation of sample designs for estimating a locally concentrated pollutant. Commun Stat Sim Comput 16:735–759

McBratney RW, Burgess TM 1981a The design of optimal sampling schemes for local estimation and mapping of regionalized variables. I. Comput Geosci 7:331–334

McBratney RW, Burgess TM 1981b The design of optimal sampling schemes for local estimation and mapping of regionalized variables. II. Comput Geosci 7:335–336

Roesch FA Jr 1993 Adaptive cluster sampling for forest inventories. Forest Sci 39:655–669

Rubin DB 1976 Inference and missing data. Biometrika 63:581–592

Scott AJ 1977 On the problem of randomization in survey sampling. Sankhya 39:1–9

Smith DR, Conroy MJ, Brakhage DH 1995 Efficiency of adaptive cluster sampling for estimating density of wintering waterfowl. Biometrics 51:777–778

Solomon H, Zacks S 1970 Optimal design of sampling from finite populations: a critical review and indication of new research areas. J Am Stat Assoc 65:653–677

Sugden RA, Smith TMF 1984 Ignorable and informative designs in survey sampling inference. Biometrika 71:495–506

Thompson SK 1988 Adaptive sampling. Proceedings of the Section on Survey Research Methods of the American Statistical Association. American Statistical Association, Alexandria, VA, p 784–786

Thompson SK 1990 Adaptive cluster sampling. J Am Stat Assoc 85:1050–1059

Thompson SK 1991a Adaptive cluster sampling: designs with primary and secondary units. Biometrics 47:1103–1115

Thompson SK 1991b Stratified adaptive cluster sampling. Biometrika 78:389–397

Thompson SK 1992 Sampling. Wiley, New York

Thompson SK 1993 Multivariate aspects of adaptive cluster sampling. In: Patil GP, Rao CR (eds) Multivariate environmental statistics. Elsevier, New York, p 561–572

Thompson SK 1996 Adaptive cluster sampling based on order statistics. Environmetrics 7:123–133

Thompson SK, Ramsey FL 1983 Adaptive sampling of animal populations. Department of Statistics, Oregon State University, Corvallis, OR (Technical Report 82)

Thompson SK, Seber GAF 1994 Detectability in conventional and adaptive sampling. Biometrics 50:712–724

Thompson SK, Seber GAF 1996 Adaptive sampling. Wiley, New York

Thompson SK, Ramsey FL, Seber GAF 1992 An adaptive procedure for sampling animal populations. Biometrics 48:1195–1199

Yfantis EA, Flatman GT, Behar JV 1987 Efficiency of kriging estimation for square triangular and hexagonal grids. Math Geol 19:183–205

Zacks S 1969 Bayes sequential designs of fixed size samples from finite populations. J Am Stat Assoc 64:1342–1349

DISCUSSION

Barnett: I welcome this contribution. The whole question of how data are collected is vital to all aspects of smart farming. In particular, newer techniques, such as the adaptive sampling Steven Thompson has explained to us, have enormous importance. We can get massive increases in efficiency by using techniques of this sort. More and more, people in this field have got to become aware of these methods.

I would like to mention a different sampling method which is within the same category, and very much in the spirit of the fact that smart farming means smart sampling, and smart sampling needs smart samplers. Suppose that I wanted to sample a field for an average level of some chemical, for instance, and have little money available so I am going to take a small sample of five observations. I could pick five sample points at random and I could work out a sample mean, or I could be a bit smarter and ask the farm owner where I would get the highest value, and then ask the farm manager where the smallest value would be found, and so on. In this way I choose five points, and if I work out the average of those I would get something like a 200–300% gain in efficiency, so those five observations would be rather like, in value, 50 observations in a random sample. You might ask me, 'Suppose these people didn't know what they were talking about it?' It does not matter, you cannot be any worse off — in this case it is as good as a random sample, and if they have any knowledge at all it is going to be better. This is a method of sampling called 'ranked set' sampling, which does not seem to be well known in the agricultural world. Intriguingly, it was discovered by an agriculturist and published in an agricultural journal, namely the *Australian Journal of Agriculture* in 1952 by McIntyre in the context of estimating pasture yields. It has moved on a long way since. This illustrates two points: first, non-random sampling is crucial in the sort of problems we are talking about; and second, we ought to make use of expert knowledge.

Groffman: This technique sounds like a disaster in terms of bias.

Barnett: No, it is totally unbiased.

Groffman: It depends on what variable you are sampling and what the knowledge base is. If we think we have a field that is contaminated with some chemical, are we then going to ask the owner of the field which areas are going to have the highest and lowest levels and expect him to give us unbiased objective answers?

Barnett: You do not ask any one person to give you more than one selective judgement. Each of the points is on a different selective judgement basis. It makes no difference if these people are totally uninformed: you cannot do worse than in random sampling, and if they have any information at all you do better. This technique is used widely in regulatory work. I have illustrated it in a deliberately rather 'folksy' way: more often the choice is not subjective, it is based on objective knowledge.

Rasch: Steven Thompson, I found your paper impressive, and the sampling concept Vic Barnett has described has some connection with your presentation. You have to have some prior knowledge for your sampling: either you get it from the farmer, or you get it by a first stage (step) of the two-step adaptive sampling procedure. Random sampling is not a good procedure because as a random procedure it cannot be improved. A two-stage procedure gives you some prior information in the first stage and then this information will be used for the second step. The question is: is a two-step procedure the best, or should we have a third stage? What is the optimal number of stages and the optimal total size of the observation with respect to the precision we need? One possibility for defining the precision of sampling is to use the mean squares, but you can also look for the half expected width of a confidence interval we have to construct for our predictions.

Thompson: Another example of an objective other than mean square error is when people are sampling soil for environmental remediation. If it's contaminated they have to do something very expensive, so they really want to find the boundary between the area in which remediation is required and the area not requiring remediation. An initial sample is observed and then additional sample sites are added near that boundary, that is, near the initial sites with contamination levels close to the threshold level for remediation.

Rasch: One can think of many optimality criteria: you get of course for each criterion an optimum design. In this sequential procedure, you showed us that the efficiency is increasing. Has this something to do with the fact that the average sample size in sequential procedures is smaller than for fixed sample size (one-stage) procedures? You compared a fixed sample size with a sequential or two-step sample size.

Thompson: But I used the same average sample size as my basis of comparison. The adaptive sampling would look still more efficient if you took into account the cost, because it may be less expensive to take a cluster of samples than to take the same number of samples evenly spread across a larger area.

Rasch: But for the same variance one would need a smaller average sample size in the adaptive procedure. Does this have something to do with the average sample size in sequential sampling?

Thompson: Yes. In fact, adaptive sampling is a type of sequential procedure. But the possibilities go well beyond what is usually considered in the field of sequential analysis. 'Sequential sampling' is usually defined more narrowly to mean a data-dependent stopping rule in collecting independent observations. With the types of adaptive designs we are talking about, one can look sequentially at the observations as they are obtained and choose exactly where to make the next observations, not just how many to make. And the design-unbiasedness of the estimates does not depend on any assumption about independence. Actually, we are taking advantage of the high spatial dependence when we look for more insects nearby when high abundance is observed.

Burrough: Adaptive sampling and progressive sampling have been used in photogrammetry for 25–30 years for making digital elevation models from stereo aerial photographs, using exactly the procedure which you have explained (e.g. Burrough 1986).

One of the most extreme kinds of samples you can get is when people digitize contour lines in order to create raster digital elevation models. If you take your contour samples, you have sets of data which all have the same z value, and they are very strange data. If you try to interpolate those with kriging where your search circle is such that it only picks up data from one contour line, you can turn nice smooth surfaces into sets of terraces. This is not appreciated by many people and you get it in standard packages such as ARC-INFO if you use kriging interpolation on these digitized contour lines. This is where you are getting such an extreme directed form of sampling that it doesn't satisfy any geostatistics techniques, but people misuse it. This is why other techniques are much more appropriate.

Thompson: Does the progressive sampling use stages of auxiliary information such as double or multiphase sampling?

Burrough: No, there are two things. If you work from stereo aerial photographs on a stereo plotter, you are then working from the original stereographic information so you then have a regular grid sample which is fine-tuned to take account of the detail which the operator sees, and that is now done automatically. The next thing is when somebody has the paper map which has the contour lines on it and they want to create digital elevation models to plot their crop yield data on. If they then start digitizing their contours they have all these digitized points which, if they then try to interpolate, often they will end up with nasty steps in the landscape. You can avoid this by using the correct kinds of interpolator.

Mulla: Would that problem not exist if you were to lay a grid over a map and then take the elevation at each grid point rather than along the contours?

Burrough: That would be very much better, but if you start with contours you have to interpolate by eye what the height is, thus introducing another error. Most people find it easier to run the cursor along the contour line and just read the level off. If you have terrain where the contours are close together this is not a problem because you get lots of data from different elevations in your circle, but lower down the slope where the contours are further apart you only pick up one contour line.

Su: In practice, it tends to be difficult to carry out some sample methods suggested by statisticians because of the limitations of manual labour, etc. It is easier to use random sampling.

Thompson: Most of these designs are motivated in the first place by that problem: you are faced with some spatial sampling problem where you can't overcome all your problems by a larger sample size. You have to take advantage of your spatial structure and the lack of independence to get the most value for your effort. For example, in sampling a field for eggs of an insect species that are highly clustered, so that the eggs are mostly concentrated in small patches of the field, a conventional sampling plan such as a systematic or random one will mostly find sites without any eggs. That gives the conventional estimates a high variance or a low precision. The adaptive design, adding neighbouring sites to the sample whenever a certain quantity of eggs are found, seeks to increase precision for a given amount of sampling effort by focusing a larger portion of that effort into regions of the field with the most information.

Webster: You mentioned that miners have a system of adaptive sampling whereby they start off with a coarse grid and then they intensify this where the concentrations of the deposit they are mining are large. Typically, they decluster to obtain estimates which they hope are unbiased. Would you like to expand on that? I'm not very sure of the theory. Are the estimates truly unbiased by this declustering scheme?

Thompson: Yes, the design-unbiased estimates of adaptive cluster sampling are truly unbiased, regardless of what the population is like, because of the way the sample is selected. The weighting of the observations that gives the unbiasedness is based on the fact that sites within the natural clusters have a higher probability of being included in the sample than the other sites, so the weighting compensates for the unequal probabilities. Model-based estimates such as kriging estimates, on the other hand, use weights based on assumptions about the characteristics of the population, in particular about the spatial covariance structure which describes the tendency of nearby sites to be relatively more similar than distant sites. In the mining example, the kriging estimate was unbiased with an assumed Gaussian (normal) spatial model if the sample was selected by a conventional design such as a systematic one. The same estimate was not unbiased when the sites were selected adaptively as the mining company did, but the bias should be relatively small because the kriging estimate would still have maximum likelihood properties whether the design was adaptive or not.

Webster: It also involves estimating the variogram in the case of the kriging, and presumably one must decluster or down-weight the points in the more densely sampled parts of the region when doing that.

Gómez-Hernández: The most commonly used declustering technique is simple cell declustering. The area under study is partitioned into a number of cells. Then each datum receives a weight proportional to the inverse of the number of data in the cell. The mean value is obtained with these weights. The process is repeated for different cell sizes and the weights selected are those corresponding to the smallest mean estimate, if the data are clustered in the highs, or the highest mean estimate, if the data are clustered in the lows. Another method for declustering is block kriging.

Thompson: The declustering procedure you describe is actually very similar in spirit to the simplest of the design unbiased estimates with adaptive cluster sampling.

GIS support for precision agriculture: problems and possibilities

Arnold K. Bregt

DLO Winand Staring Centre for Integrated Land, Soil and Water Research (SC-DLO), PO Box 125, 6700 AC Wageningen, The Netherlands

Abstract. Precision farming aims to optimize the use of soil resources and external inputs on a site-specific basis. Base ingredients for research in the field of precision farming are spatial data, including a characterization of the spatial variability, and simulation models for the characterization of the processes that take place. Geographical information systems (GIS) are systems for the storage, analysis and presentation of spatial data. A combination of GIS and simulation models is highly relevant for precision farming. Currently only static one- or two-dimensional simulation models can be fully supported by commercial GIS systems. Within precision agriculture an engineering component can be also distinguished, in which the research findings are translated into operational systems for use at farm level. GIS can support this engineering activity by providing a good platform for storage of base data, simple modelling, presentation of results, development of a user interface and, in combination with a global positioning system, controlling the navigation of farm vehicles. On the basis of GIS a decision support system could be developed for operational application of precision agriculture at farm level.

1997 Precision agriculture: spatial and temporal variability of environmental quality. Wiley, Chichester (Ciba Foundation Symposium 210) p 173–181

Agricultural practices are changing in response to economic, technological, environmental and social trends. The profitability and environmental impact of farming systems are of major concern for both farmers and the government.

Precision agriculture, which takes into account spatial variation of soil at field level, is being proposed as a remedy to many of these environmental problems (Larson & Robert 1991). The concept behind it is based on matching the external inputs, such as fertilizers and herbicides, with the variation in local soil conditions, in order to reduce costs and negative environmental effects.

Over the last 10 years there have been many publications on precision agriculture. For example, in the proceedings of the conference on site-specific management for agricultural systems (Robert et al 1995) a number of papers are on this subject. Many research papers deal with soil variability, yield monitoring, modelling crop growth and nutrient leaching at farm level.

The research components for precision farming are information about the site-specific soil conditions and details of the processes that occur in crops and soils. Information about the soil is generally presented in the form of point-sampled data and detailed soil maps. Research on the processes that occur in crops and soils is condensed in the form of simulation models. Besides a research component, precision agriculture also has a strong engineering component. One of the goals of research in precision agriculture is to produce operational systems which can be used at farm level.

The use of geographical information systems (GIS) for the storage, management, analysis and presentation of spatial data has diversified rapidly over the last 10 years. GIS applications have been developed for a variety of fields, ranging from land use planning and utility management at local scale to global warming and acid deposition on a global scale. The number of publications dealing with the use of GIS in precision agriculture are, however, limited. Han et al (1995) describe the linking of GIS with a potato simulation model for site-specific crop management. Griffith (1995) propagates the use of GIS for economic analysis at farm level. In most papers on the use of GIS in precision agriculture the function of GIS is limited to data pre-processing to produce field management maps (Long et al 1995) or to generate (some of) the input data for simulation models.

This paper discusses the problems and possibilities for GIS support in research and engineering in the field of precision agriculture.

Research in precision agriculture and GIS support

From a research viewpoint precision agriculture deals with data-related and modelling-related issues, and the interaction between data and modelling. In precision farming geographical data on various aspects such as soil, crop, weather and field history are relevant. Soil data are used in the form of point observations, describing properties of the soil in the z dimension, and soil maps describing the spatial extent of the soil pattern in the x,y dimension. Crop parameters can be considered to be constant at field level. Weather data can be considered as spatially constant at field level, but they are highly variable over time. Field history, e.g. the amount of fertilizer applied and the different crops grown, varies both in space and time. In summary, we see that geographical data are recorded in three dimensions in space and one in time, which makes data management and analysis in precision agriculture a complex business.

A second component in precision agriculture research is simulation modelling. For example, simulation models exist for the flow of water, crop growth, soil erosion, nutrient and pesticide leaching. There seems to be no general accepted classification of process models. Model types often found in literature (e.g. France & Thornley 1984, Burrough 1992) are:

Empirical and mechanistic models. An empirical model describes a process based on empiricism, whereas a mechanistic model attempts to give a description with understanding.

Static and dynamic models. A static model does not contain time as a variable: any time-dependent components of the behaviour of the system are ignored. Since all processes in the world involve change, a static model is always an approximation. It might be a good approximation if the phenomena are close to equilibrium. A dynamic model, on the other hand, contains the time variable in the equations.

Deterministic and stochastic models. A deterministic model is one that makes definite predictions for quantities (such as crop yield, rainfall), without any associated probability distribution. A stochastic model, on the other hand, contains some random elements or probability distributions. The model not only predicts the expected value of a quantity, but also the variance. The greater the uncertainty in the behaviour of the process, the more important it is for a model to follow a stochastic approach. Stochastic models tend to be technically difficult to handle and can quickly become very complex. Another approach in dealing with uncertainty is to use a combination of a deterministic model and Monte-Carlo simulation to obtain probability estimates.

Spatial dimensions modelled. We can distinguish between one-, two- and three-dimensional models.

Qualitative and quantitative models. A qualitative model makes predictions on a qualitative level, such as 'not suitable', 'suitable' or 'highly suitable'. The input for a qualitative model can be both qualitative and quantitative. A quantitative model, on the other hand, produces quantitative output.

Models constructed for real-world processes often contain combinations of the model types described above. For example, we may have a dynamic/deterministic/quantitative one-dimensional model for describing crop growth (Van Diepen et al 1989) or a static/empirical/qualitative one-dimensional model for land evaluation (Van Lanen 1991). For the integration with GIS the characteristics 'dimensions' and 'static -dynamic' are of major importance. Most of the models applied in precision agriculture are dynamic one- or two-dimensional models.

Current GIS can support the research (modelling) in precision agriculture in the following way:

(1) The data collected can be stored and pre-processed in GIS in order to generate spatially differentiated input data for the simulation models. As shown by Bregt (1993) this can only be done fully for static one- or two-dimensional simulation models. The main reason for this is that the data models underlying commercial GIS are based on a two-dimensional description of the Earth's surface. For full GIS support of the dynamic simulation models (the most relevant models used in precision agriculture) the geographical data model underlying commercial GIS must be expanded in the time dimension. Although some research prototypes

have been produced, this has not yet resulted in an commercially available product.

(2) For simple one- and two-dimensional models the simulation model can be implemented fully in the GIS environment, using the analysis functions of the GIS software (tight integration). The advantage of this approach is that a more integrated system for data storage, modelling and presentation is created. A disadvantage is that this limits us to relatively simple models.

For simulation models in precision agriculture, loose integration (Stuart & Stocks 1993) is currently the best option. In the loose-coupled approach the simulation model and GIS are linked loosely through an interface. The interface may consist of simple manual transfer of ASCII data files or the development of programmes which ease the data transfer. The loose coupling is flexible and many models can be integrated. The choice of a particular type of integration (loose or tight) depends on the situation at hand. In the case of a complicated dynamic process model, loose integration is presently the only practical option. Also the large investment in encoding process models combined with the limited functionality of GIS to implement these implies that a loose coupling to GIS is the best solution for the more advanced applications. In the case of relatively simple models a tight integration with GIS is recommended as these applications are easier to develop and maintain, and allow for easier animation of the model results.

(3) The presentation of spatial data in the form of maps is one of the most appealing features of GIS. This function of GIS can be used to present the modelling results in a form which can be easily communicated to others.

Engineering in precision agriculture and GIS support

One of the main goals of research in precision agriculture is that it will lead to operational systems for use at farm level. A few operational systems are described in the literature. Robert (1989) described a computerized spreader for fertilizers and chemicals as a function of local variation in soil conditions. McGrath et al (1995) described an expert system for combining soil test results with data concerning crop models, fertilizers, chemicals, weather, field history, seed characteristics and yield economics. The use of GIS in the reported systems is, however, limited. There are a number of reasons for this:

(1) Precision agriculture is a relatively new research area which has not yet received the full attention of the more applied research organizations, who are more focused on the engineering side.
(2) The structure of data used in precision agriculture is quite complex when compared with the data models supported by current commercial GIS.

(3) Commercial GIS systems are still quite expensive for operational use within an integrated system for precision agriculture.
(4) There are still limited digital spatial data (soil and topography) with sufficient resolution to be used at farm level.

Although the present use of GIS for precision agriculture is limited, there is good potential in the near future for the development of a GIS-based decision support system (DSS) at farm level. What are the requirements of such a system, and how can it be constructed using GIS technology? It must be weather proof, user-friendly, fast, interactive, simple, integrated with GPS, and it must contain spatial data on soil and relationships between soils, crops, production, and nutrient and pesticide leaching as a function of the input of fertilizers and farm chemicals.

For the construction of such a DSS, the keyword is simplicity. If we try to condense all our research findings into one system it will be far too complicated to be used at farm level. A brief description of the main components of such a system follows.

Spatial database

The spatial database must contain a two-dimensional spatial representation of the farm. The minimum dataset to be included in such a database is a detailed topographic map of the farm and a spatial characterization of the soil in the form of a detailed soil map with representative profiles and associated attributes.

Knowledge module

This module stores details of simple relationships between e.g. soil/groundwater fertilizer application and nutrient leaching, fertilizer application and crop production, and input–output relations for economic evaluation. The knowledge module should not contain complex simulation models but simple relations. These relations could be derived by using: (a) expert knowledge which is formalized in the form of tables and expert rules, and (b) dynamic simulation models for modelling various scenarios of soil/water/crop relations for various input levels and summarizing the modelling results in the form of meta-models.

A distinction should be made between non-spatial (e.g. economic relations) and spatial relations (e.g. nutrient leaching as a function of soil type). Most of the non-spatial relations are probably already implemented in the existing management support systems for farmers. The spatial relations must be derived for the specific physical conditions of the farm.

Locator module

In order to apply precision agriculture, it is essential to relate the data in the database with the position on the land. Recently a number of GIS products (e.g. Locator GIS) have been released which provide this as standard functionality.

User interface

A user-friendly interface is vital for the operational use of the system.

The above described DSS could currently be developed by the use of state-of-the-art GIS technology in combination with user interface tools. The new generation of GIS systems are, from a software point of view, more open (or are becoming more open) than the old systems, which mean that they can be integrated with programming environments such as Delphi or Visual Basic. It is now time to construct a GIS-based DSS for precision farming on the basis of the research findings.

Conclusions and recommendations

The integration of GIS and process models offers interesting possibilities for enlarging the analysis for precision farming. GIS forms a good platform for the storage and management of model input data and the presentation of model results, while the process model provides the analysis capabilities which are lacking in current GIS. If we, however, confront the current data models underlying commercial GIS with the data requirements of the simulation models, it appears that only simple, static one- or two-dimensional models can be integrated easily. For the dynamic process models, which from an analysis viewpoint are the most interesting ones for precision farming, GIS can only partly play a role in storage and management of model input data and presentation of results. In order to increase the integration possibilities, the geographical data model underlying commercial GIS must be expanded with the time dimension.

The prospects for creating an operational DSS for precision farming by using GIS technology are good. A new generation of GIS products are appearing on the market, with built in GPS facilities and a more open structure, which makes the development of a DSS for precision agriculture on the basis of the existing non-spatial management support systems possible. Such a DSS should at least contain the components database, knowledge module, locator module and a user interface. As a first step the knowledge module should contain simple relations in form of meta-models, knowledge tables and simple simulation models. It is a challenge for the research community to derive these relations from the more complex dynamic simulation models, which are used for research in the field of precision farming.

References

Bregt AK 1993 Integrating GIS and process models for global environmental assessment. In: Tateishi R (ed) Proceedings of the international workshop on global GIS, August 24–25, ISPRS, Tokyo, Japan, p 77–84

Burrough PA 1992 Development of intelligent geographical information systems. Int J Geogr Info Sys 6:1–11

France J, Thornley JHM 1984 Mathematical models in agriculture. Butterworth, London

Griffith D 1995 Incorporating economic analysis into on-farm GIS. In: Robert PC, Rust RH, Larson WE (eds) Site-specific management for agricultural systems. ASA, CSSA, SSSA, Madison, WI (Proc 2nd Conf, Minneapolis, 1994), p 723–729

Han S, Evans RG, Hodges T 1995 Linking a geographic information system with a potato simulation model for site-specific crop management. J Environ Qual 24:772–777

Larson WE, Robert PC 1991 Farming by soil. In: Lal R, Pierce FJ (eds) Soil management for sustainability. Soil and Water Conserv Soc, Ankeny, IA, p 103–112

Long DS, Carlson GR, DeGloria SD 1995 Quality of field management maps. In: Robert PC, Rust RH, Larson WE (eds) Site-specific management for agricultural systems. ASA, CSSA, SSSA, Madison, WI (Proc 2nd Conf, Minneapolis, 1994) p 251–271

McGrath DE, Skotnikov AV, Bobrov VA 1995 A site-specific expert system with supporting equipment for crop management. In: Robert PC, Rust RH, Larson WE (eds) Site-specific management for agricultural systems. ASA, CSSA, SSSA, Madison, WI (Proc 2nd Conf, Minneapolis, 1994) p 619–635

Robert PC 1989 Land evaluation at farm level using soil survey information systems. In: Bouma J, Bregt AK (eds) Land qualities in space and time. Pudoc, Wageningen (Proc Symp, Wageningen, 1988) p 299–311

Robert PC, Rust RH, Larson WE (eds) 1995 Site-specific management for agricultural systems. ASA, CSSA, SSSA, Madison, WI (Proc 2nd Conf, Minneapolis, 1994)

Stuart N, Stocks C 1993 Hydrological modelling within GIS: an integrated approach. In: HydroGIS 93: application of geographic information systems in hydrology and water resources. IAHS Publication 211, p 333–343

Van Diepen CA, Wolf J, Van Keulen H, Rappoldt C 1989 WOFOST: a simulation model for crop production. Soil Use Manag 5:16–24

Van Lanen HAJ 1991 Qualitative and quantitative land evaluation: an operational approach. PhD Thesis, Agricultural University, Wageningen, The Netherlands

DISCUSSION

Robert: There are currently at least 10 US companies developing commercial GIS for precision agriculture. Some systems will have most of the characteristics you mentioned—and certainly the user-friendliness. Today's systems are either limited in function, unstable, or difficult to operate, but this is improving continuously. To my knowledge, decision support systems are not yet available.

Rabbinge: As has been made clear in the discussions we have had so far, although there is a strong technology push for precision agriculture, its scientific basis needs further consideration. There are still many blind spots in our knowledge and, in order to fill these in, we need more research. When we introduced supervised pest and disease control systems in the early 1970s there was similar hype, but much of the interest faded away because there was no continuous upgrading of the systems and their scientific basis was limited. It took more than a decade to get the same commitment and willingness from the farmers for their application again. If we are not working on these aspects, precision agriculture may suffer a similar fate.

Goense: Arnold Bregt mentioned that for modelling at least four dimensions (x, y, z and t) were required. I tried to do some meta-model development including spatial attributes. I think there's a fifth dimension required: how were the data obtained?

Are they measured or simulated data? If they are predicted data, I want to know which type of model predicted them. If they are measured data, which type of technology was used and which calibration tables were used? There are many additional attributes relevant to those objects, which makes me a little sceptical about the use of presently available GIS, even when the fourth dimension of time is included.

Burrough: You are making two points there: one concerns the data quality (where do they come from, what is their provenance and how are they collected?) and the other concerns whether you have the data models, the data structure and the algorithms to be able to do the processing that you want. Arnold Bregt highlighted commercial GIS, which has a particular cartographic, static data model. It is very successful in various areas, but it's not particularly successful if you're dealing with processes of change in space and time. What you find is that people who are interested in space and time problems are moving away from commercial GIS (unless their institute forces them to use it) to develop other methods. People are moving away from the commercial systems to more specialized ones, providing systems that the market wants, because it's no longer a big deal to develop them: you can put a system together much more cheaply today than you could have done five years ago. The future market for GIS is going to change radically over the next few years, with availability of data and system-interoperability, and this is going to come back to the business of data quality.

Rabbinge: That's an encouraging message.

Burrough: Systems are also getting easier to use. It used to take a graduate six months to learn how to use ARC-INFO, whereas new systems can be learned in a few days.

Bregt: I partly agree with you. You are right that the whole GIS market is breaking up. It is being taken over by specialized organizations who are developing components of GIS infrastructure, rather than developing complete GIS. This hasn't happened fully yet: the commercial systems are still dominating the market, but I expect that in a few years the whole scene will be completely different.

Stein: We have had some discussion today about the quality of models. If you run your models together with GIS you are going to be at risk of always getting a result. This might be a danger, in particular with the aim and the appropriate scale of the models. You didn't raise this issue in your talk: did you avoid it deliberately?

Bregt: I didn't intend to say anything about this because I didn't concentrate on a particular model or model quality, but just its general characteristics. We have integrated GIS quite a lot with models. What you see to a certain extent is dangerous because you're building more complex systems, but it's also helping you to see what's wrong, because presenting data in the form of a map is a powerful medium for detecting errors. If you are modelling and you are presenting the information and using GIS as the backbone of your modelling exercise this is making you more aware of what you're doing.

Stein: If you go with the simplicity you run the risk that very simple models can always be applied with approximate data sets. This risk is much larger if you go with more complexity. Wouldn't it be an essential tool for GIS to have some warning about the level of complexity for the particular model, e.g. when it could go wrong?

Rabbinge: It's wise to incorporate this sort of danger signal. We've tried to do this in many of the simulation models. For example, in many models a dimension checker and an energy and carbon balance is needed so that that you are not generating or losing any carbon or energy.

Bouma: I have a comment on the models. You said some important things about this, and also about the ultimate system where operations must be relatively simple. We are getting some results on this. We used the validated model for our particular situation in terms of crop growth and nitrogen cycling. We had measured water table values. For a given dataset, you can run a lot of realizations and they can together be generalized by the kind of equations that you talk about in terms of a meta-model. This is a powerful approach.

In a second set of experiments we looked at a set of fields with different fertilization rates using GIS to visualize results. We found that a rather wide range of fertilization rates didn't have much effect on yields and leaching but then all of a sudden there are these threshold values where conditions really change. I have a feeling that those kinds of threshold values can be tied up with soil types. We don't hear much about soil survey any more. I'm convinced that we can use simulation to feed the meta-models but, more importantly perhaps, we can define threshold values for the various soil types where behaviour really changes. Perhaps putting these in the system could be a helpful step towards the systems that Arnold Bregt has talked about.

Modelling for precision weed management

M. J. Kropff*†, J. Wallinga† and L. A. P. Lotz†

**Department of Theoretical Production Ecology, Wageningen Agricultural University, PO Box 430, 6700 AK Wageningen, and †DLO Research Institute for Agrobiology and Soil Fertility (AB-DLO) PO Box 14, 6700 AA, Wageningen, The Netherlands*

Abstract. Recently, the need for the development of weed management systems with a reduced dependency on herbicides has increased because of concern about environmental side-effects and cost. The development of such systems requires new strategies based on improvements with respect to (1) prevention, (2) decision making and (3) weed control technology. For the development of improvements in all three aspects, quantitative understanding of weed population dynamics and crop–weed interactions is needed. Models that integrate the available quantitative knowledge can be used to design preventive measures, to develop long-term and short-term strategies for weed management, to assist in decision making to determine if, when, where and how weeds should be controlled and to identify new opportunities for weed control. Ecophysiological simulation models for crop–weed competition simulate growth and production of species in mixtures, based on ecophysiological processes in plants and their response to the environment. Such models help improve insight into the crop–weed system and can be used for purposes such as the development of simple predictive yield-loss models, threshold levels or the design of competitive crop plant types. For strategic weed management decisions, preventive measures and the identification of new opportunities for weed control, quantitative insight into the dynamics and spatial patterns of weed populations is also required. The complexity of the process and the long-term character of weed population dynamics make the use of models necessary. Different modelling approaches have been developed and are described briefly. Opportunities to use the available knowledge and models to improve weed management are discussed. Weeds occur in patches and their sensitivity to herbicides changes strongly with developmental stage, making precision techniques for herbicide application in time and space an option for reducing herbicide use. Limitations related to insight in biological processes as well as the state of technological development are discussed.

1997 Precision agriculture: spatial and temporal variability of environmental quality. Wiley, Chichester (Ciba Foundation Symposium 210) p 182–204

In most agricultural systems weed management has been one of the key issues, especially before herbicides became available. The use and application of herbicides has been one of the main factors enabling the intensification of agriculture that has occurred in developed countries in the past decades. More recently, the availability of herbicides has been coupled to intensification of agriculture in developing countries as well. However, increased concern about environmental side-effects of herbicides, the development of herbicide resistance in weeds and the necessity to reduce the cost of inputs have resulted in greater pressure on farmers to reduce herbicide use. This has led to the development of new strategies for weed management.

The strategy for improving weed management systems on the basis of increased precision consists of three components (Kropff 1996):

(1) Reducing weed effects through adapted crop management (*prevention*).
(2) Improving decision making with respect to weed control (*decision making*).
(3) Improving control technology (*control*).

The first component (*prevention*) involves any aspect of management that favours the crop relative to the weeds. This includes the development of competitive crop varieties with minimum trade-off between competitiveness and yield potential of the cultivar. Systems analysis and simulation are indispensable tools for the study of such complex interrelationships and may help bridge the gap between knowledge at the process level and management at the field level.

The second component (*decision making*) consists of strategic (long-term) decisions, tactical decisions (for a season) and operational decisions in the field. This requires long-term and short-term strategies for weed management, to assist in decision making to determine if, when, where and how weeds should be controlled.

The third component (*control*) deals with the development/improvement of weed control technology and is strongly related to precision technology. Three methods of weed control can be distinguished: biological, mechanical and chemical. There are many ways in which control technology can be improved ranging from precision mechanical weed management tools to precision herbicide treatments.

The improvement of weed management systems, on the basis of this three-component strategy, requires quantitative insight into both crop–weed interactions and the dynamics of weed populations in space and in time. Because of the complexity of the processes and the long-term aspects in population dynamics, models are required to obtain such quantitative insight and to make the knowledge operational.

This paper reviews current quantitative knowledge on crop–weed interactions and weed population dynamics and discusses possibilities for using this knowledge to develop precision weed management systems on the basis of the three components mentioned above.

Modelling crop–weed interactions and weed population dynamics

Modelling crop–weed interactions

Weeds affect crop yield as a result of competition between the crop and weeds for the growth-limiting resources of light, water and nutrients. The quantitative understanding of crop–weed interactions has been reviewed in detail by Kropff & Van Laar (1993). Two types of models for crop–weed interactions have been developed: (1) descriptive regression models with a few parameters that can be determined by fitting the model to observed data, and (2) expanded ecophysiological crop growth models that simulate competition for the growth-determining (light) and growth-limiting (water and nutrients) resources between species.

Descriptive models for crop–weed interactions

The most widely used regression model for describing the effects of competition at a certain moment is the hyperbolic yield loss weed density model (Cousens 1985):

$$Y_L = \frac{aN_w}{1 + (a/m)N_w} \tag{1}$$

where Y_L gives the yield loss, N_w is the weed density, a describes the yield loss caused by adding the first weed per m^2 and m the maximum yield loss. These hyperbolic yield density equations fit well to data from experiments where only the weed density is varied (Kropff et al 1984, Cousens 1985). However, the parameters a and m may vary strongly among experiments due to the effect of other factors on competition processes (Kropff & Van Laar 1993). Because both weed density and the period between crop and weed emergence strongly determine the competitive relationship between crops and weeds (Kropff et al 1984, Cousens et al 1987, Kropff et al 1992), more robust prediction of yield loss on the basis of early observations should be based on these two factors.

A simple descriptive regression model for early prediction of crop losses by weed competition introduced by Kropff & Spitters (1991) and extended by Kropff et al (1995) was derived from the well tested hyperbolic yield density model. This model relates yield loss to relative weed leaf area (L_w expressed as weed leaf area /crop+weed leaf area) shortly after crop emergence, using the 'relative damage coefficient' q as the main model parameter next to the maximum yield loss m:

$$Y_L = \frac{qL_w}{1 + ((q/m) - 1)L_w} \tag{2}$$

Because leaf area accounts for density and age of the weeds, this regression model accounts for the effect of weed density and the effect of the time of weed emergence (Kropff & Spitters 1991). An example is given in Fig. 1 which shows the relation between yield loss in sugar beet and the parameters density and relative leaf area of *Chenopodium album* L. based on data from Kropff et al (1995). Lotz et al (1996) evaluated the approach over a wide geographic region and found that the descriptive value of the model is good, but that its current predictive ability is still insufficient for precision weed management.

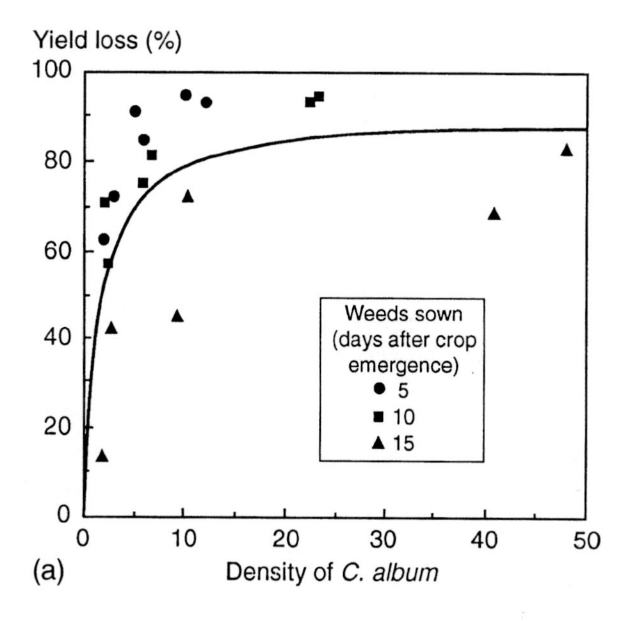

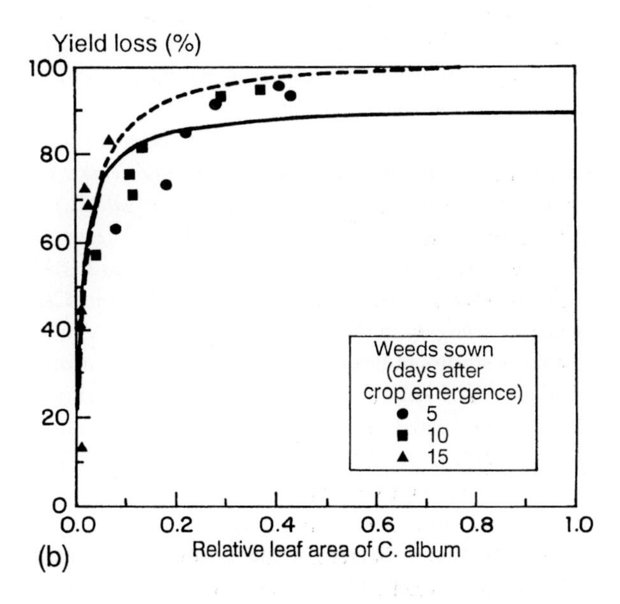

FIG. 1. Yield loss in sugar beet related to (a) the density and (b) the relative leaf area of *Chenopodium album* L. for different dates of weed emergence: circles, 5 d; squares, 10 d; and triangles, 15 d after the crop. Redrawn after Kropff et al (1995).

In conclusion, the relative leaf area–yield loss regression model accounts for the effect of weed densities, different flushes of weeds and the period between crop and weed emergence. However, the effect of other factors, such as transplanting shock or severe water stress, is not accounted for, because the regression models do not account for underlying processes.

Ecophysiological models for crop–weed interactions

Competition is a dynamic process that can be understood from the distribution of the growth-determining (light) or growth-limiting (water and nutrients) resources over the competing species and the efficiency with which each species uses the resources. Ecophysiological models that simulate the physiological, morphological and phenological processes provide insight into competition effects observed in (field) experiments and may aid in seeking ways to manipulate competitive relations, such as those between crop and weeds by determining the most important factors in crop–weed competition.

Various competition models have been developed (Spitters & Aerts 1983, Kropff et al 1984, Graf et al 1990, Wilkerson et al 1990, Kropff & Van Laar 1993). The ecophysiological model INTERCOM described by Kropff & Van Laar (1993) consists of a number of coupled crop growth models equal to the number of competing species. Under favourable growth conditions, light is the main factor determining the growth rate of the crop and its associated weeds.

Location-specific input requirements of the ecophysiological model include geographical latitude, standard daily weather data, soil physical properties, dates of crop and weed emergence, and weed density.

The ecophysiological competition model has been tested with data from competition experiments with several datasets: maize (*Zea mays* L.), yellow mustard (*Sinapis arvensis* L.) and barnyard grass (*Echinochloa crus-galli* L.) in the Netherlands (Kropff et al 1984, Spitters 1984, Spitters & Aerts 1983, Weaver et al 1992); tomato (*Lycopersicon esculentum* L.) pigweed (*Amaranthus retroflexus* L.) and tomato eastern black nightshade (*Solanum americana*) (Weaver et al 1987) in Canada (Kropff et al 1992); and rice and *E. crus-galli* in the Philippines (Kropff & Van Laar 1993) (Fig. 2).

The results of these studies indicate that inter-plant competition for light and water can be explained on the basis of the underlying physiological processes. Several approaches to introduce spatial variability in the models are underway e.g. for row crops (B. J. Schnieders & L. A. P. Lotz, unpublished results). The main gaps in our knowledge are related to morphological development and especially the phenotypic plasticity of weeds with respect to morphological features such as height and leaf area. Applications of these models include the development of new simple predictive models for yield loss due to weeds, the analysis and extrapolation of experimental data, the analysis of the impact of sub-lethal control measures (such as low-dosages of herbicides), the design of new competitive crop plant types for weed suppression and risk analysis for the development of weed management strategies.

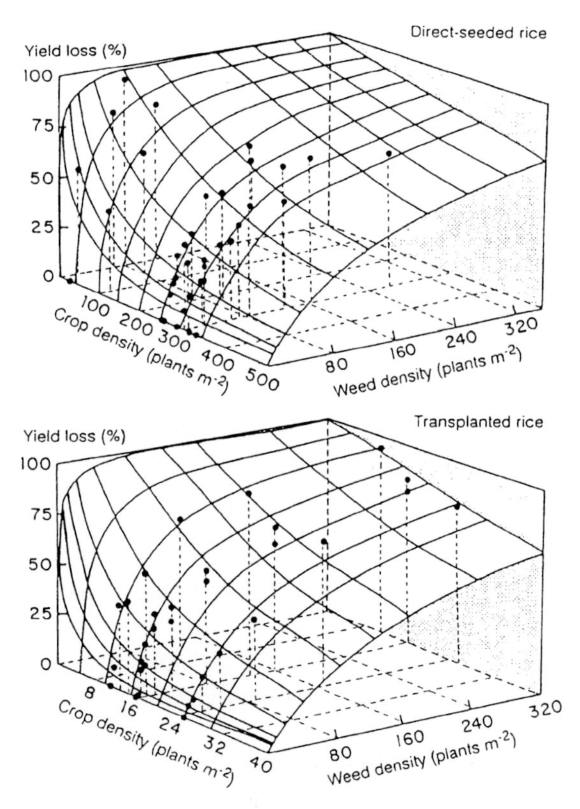

FIG. 2. Simulated (using the model INTERCOM) and observed yield loss in direct-seeded and transplanted rice caused by competition with *Echinochloa crus-galli* L. for different data sets. Redrawn after Kropff & Van Laar (1993).

Modelling temporal and spatial dynamics of weed populations

Models have been developed to integrate the knowledge on life-cycle processes. Figure 3 shows the life cycle of annual weeds. The main processes involved are germination and emergence of seedlings from the seed bank in the soil, establishment and growth of the weed plants, seed production, seed shedding and seed mortality in the soil. Competition plays a major role in establishment and growth and therefore strongly affects the population dynamics of weeds. Besides natural processes, humans have a major impact on the spread of weeds at each scale. The different mechanisms of dispersal have been discussed in detail by Cousens & Mortimer (1995) and Rew & Cussans (1995), who concluded that apart from wind dispersal few quantitative studies have been conducted on these mechanisms. Because most weed seeds remain very close to the plant (Harper 1977), weed patterns in fields do not change

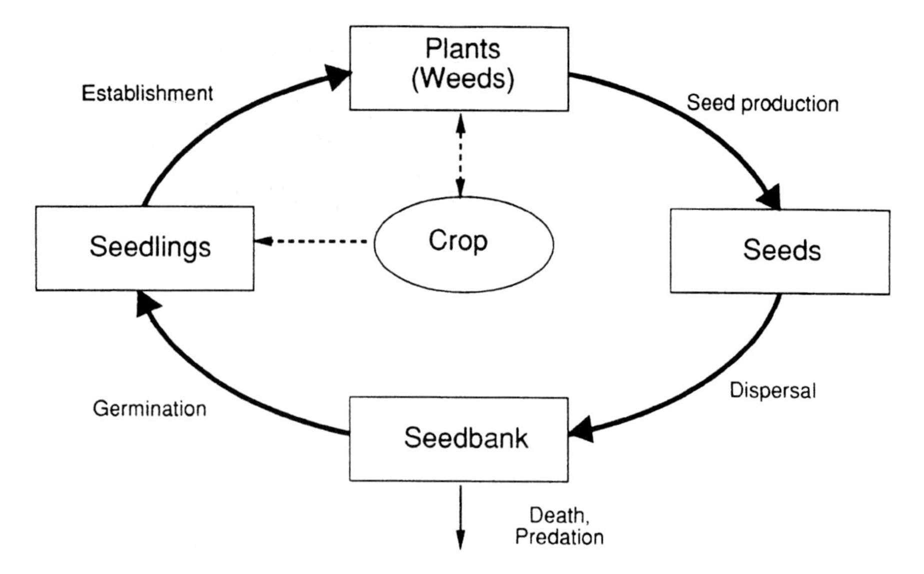

FIG. 3. Schematic diagram of the life cycle of a weed in competition with crops in annual cycles.

dramatically in time (Wilson & Brain 1991), which may provide a basis for precision agricultural practices (Johnson et al 1995).

Comprehensive models that are based on physiological principles are only available for parts of the life cycle: plant growth and competition (Kropff & Van Laar 1993), and germination and emergence (L. M. Vleeshouwers & M. J. Kropff, unpublished results). In contrast, processes such as seed shedding, seed dispersal and predation of seeds are poorly understood. The most detailed models that encompass the whole life-cycle have been developed for species like *Avena fatua* L. (Cousens et al 1986), *Alopecurus myosuroides* Huds. (Doyle et al 1986) and *Galium aparine* L. (Van der Weide & Van Groenendael 1990). The basic structure of most models has been described by Spitters (1989) and Kropff et al (1996).

Apart from the level of detail at which the life-cycle is studied, three different modelling approaches to integrate individuals into a population can be distinguished (Durrett & Levin 1994, Kropff et al 1996): (i) the density-based models; (ii) the density-based models that take spatial gradients in density into account; and (iii) the individual-based models which also account for spatial processes.

The modelling approach that is most frequently used takes density of weeds as a key variable. From the current value of the density, the rate of change in density and values for the density in the next seasons are derived (e.g. Kropff et al 1996). A tacit assumption underlying this approach is that each weed perceives a similar environment and that the system is homogeneous. The consequence of this assumption is that it is impossible to encode dispersal of weed seeds into this type of model. Yet due to the conceptual clarity in modelling temporal changes in density this approach is widely used (e.g. Firbank & Watkinson 1986).

A rather obvious way of including dispersal of weeds is to include space in the model and allow for spatial gradients in density, which results in so-called reaction–diffusion models. Discrete versions of this type of model have been employed to simulate spread of weeds (Auld & Coote 1980, Ballaré et al 1987, Maxwell & Ghersa 1992). The key variable in this modelling approach is again the weed density. Since density is interpreted as a real variable it is easy to generate artefacts such as 0.001 plant on one square metre. This problem can be overcome by truncating low densities to integer values (Schippers et al 1993, González-Andújar & Perry 1995). Another problem is that in the course of time, spatial gradients will eventually flatten out. Therefore it is hard to explain the observed 'patchiness' of weeds by these models.

One step further along this process would be for us to abandon weed density as a basic variable in the model, and proceed with the configuration of weeds over space. This modelling approach includes model types such as the individual-based model (cf. Pacala & Silander 1985) and cellular automaton models (cf. Barkham & Hance 1982, Silvertown et al 1992). This type of model makes it possible to study the interaction between dynamics and patchiness in weeds. Wallinga (1995a) analysed the development of patchiness of weeds at realistic low densities using such an individual-based spatial model. Using simulation studies, this paper demonstrated that patchiness occurs naturally at low weed densities.

Of the modelling approaches, individual-based models are the most comprehensive, but as a result of their complexity they quickly run into computing problems and such complexity is not always required. The density-based model can be very useful for roughly exploring options for long-term weed management strategies. The individual-based models can be very helpful for identifying opportunities for site-specific weed management (Kropff et al 1996).

Modelling and precision weed management

As mentioned in the introduction, three strategies can be followed to improve weed management systems based on increased precision with respect to weed management (Kropff 1996):

(1) Reducing weed effects through adapted crop management (*prevention*).
(2) Improving decision making with respect to weed control (*decision making*).
(3) Improving control technology (biological, mechanical and chemical) (*control*).

Prevention

The first component of the strategy to improve weed management involves any aspect of management that favours the crop relative to the weeds. In traditional agricultural systems, where hand-weeding was practised intensively, cropping systems were designed to reduce weed problems as much as possible. For example, varieties were selected that suppressed weeds. However, today, varieties are only selected with respect to yielding ability and product quality. In rice systems, however, increased concern about herbicide use necessitated studies on the competitiveness of rice varieties. An ecophysiological simulation model for interplant competition was used to identify traits that determine the competitive ability of a crop. The most important traits were: the development of rapid early leaf area, tiller and height, and more horizontally oriented leaves in early growth stages (vertical ones later on because of yield potential) (Kropff & Van Laar 1993). In experiments, rice varieties that differed in these traits were evaluated with respect to their competitive ability versus a standard purple-coloured variety. The variety with all the required traits, Mahsuri, reduced the growth of the purple variety so much that all purple rice plants had died before the final harvest (M. J. Kropff, unpublished results). Detailed studies on trade-offs between different traits are underway (L. Bastiaans, T. Migo & M. J. Kropff, unpublished results).

Another example of the reduction of weed effects through adapted crop management is found in the results of an experiment with three weed species in three sugar beet cultivars (Lotz et al 1991, L. A. P. Lotz & M. J. Kropff, unpublished results). Three sugar beet cultivars with strong differences in leaf angle distribution were selected, and according to model predictions such differences should have consequences for the cultivar's potential to suppress weeds. Indeed, mortality of late-emerging weeds was 25% greater when leaves of the sugar beet exhibited a more horizontal orientation which resulted in more light absorption per unit leaf area.

Decision making

The second strategy is the improvement of the decision-making process, which consists of strategic (long-term) decisions, tactic decisions (for a season) and operational decisions in the field. Here precision in time and space is required. It involves long-term and short-term strategies for weed management, to assist in decision making to determine if, when, where and how weeds should be controlled.

Strategic weed management decisions become important when threshold weed densities for weed control are introduced. The threshold weed density has to be based on the cost-effectiveness of control in the current year (is the cost of control smaller than the loss in yield?) as well as the cost-effectiveness over several years, especially when the possibility for correcting mistakes resulting in high weed infestations is limited, as it is in organic farming systems (Kropff et al 1996).

The decision-making process for tactical and operational decisions in a weed management system based on post-emergence observations is illustrated in Fig. 4. To allow rational decision making, the severity of weed infestation shortly after crop emergence should be estimated. Criteria must be defined (i.e. the objectives and planning horizon of the farmer) to enable economic decision making. First, a simple measure for estimating the severity of weed infestation has to be defined. This permits

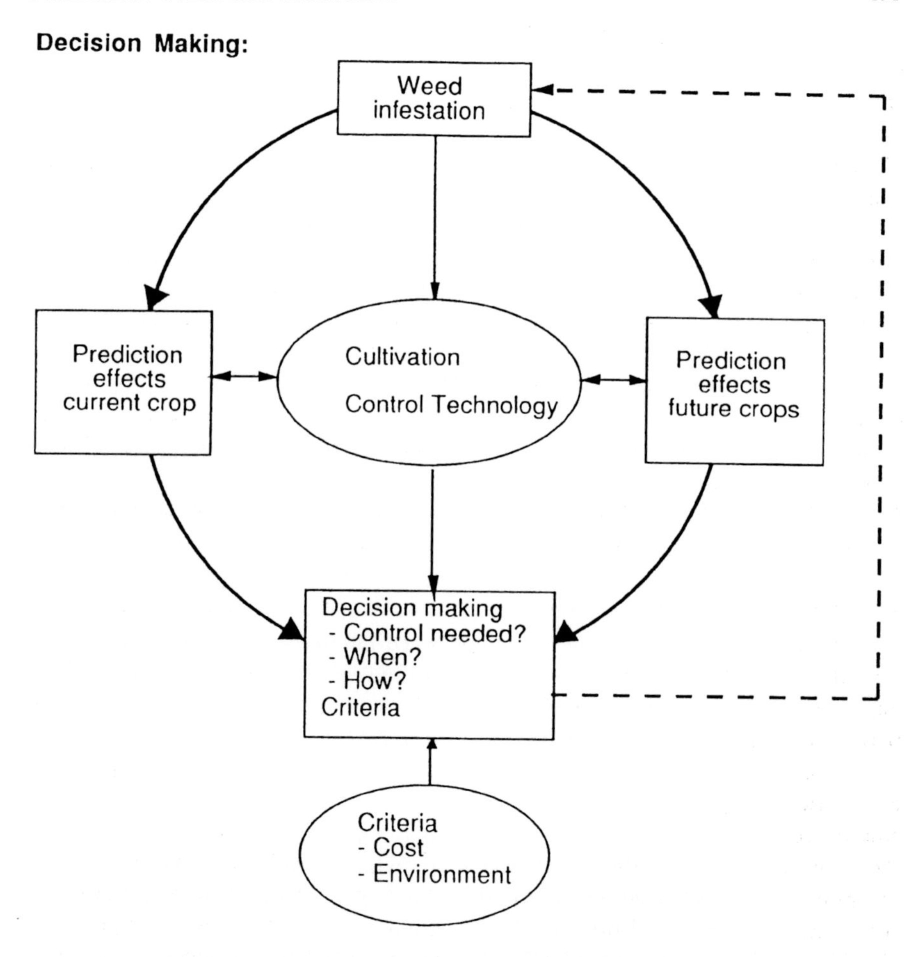

FIG. 4. Schematic representation of the decision-making process in weed management.

the prediction of crop yield loss and weed population density at the end of the growing season along with quantification of costs, efficacy and side-effects of possible control measures. This information can be used to decide whether and how the weeds should be controlled using defined criteria such as maximization of profits and minimization of environmental effects.

Is control needed? Several weed control advisory systems have been developed that use threshold densities for weed control or that focus on optimization of herbicide dosage. This threshold is the level of weed infestation which can be tolerated based on specified criteria (generally based on economic objectives) (cf. Niemann 1986, Aarts & De Visser 1985, Wahmhoff & Heitefuss 1988). Different concepts for thresholds for tactical (within season) and strategic (long-term) decision making in weed management have been developed (Cousens 1987). However, to date, the threshold approach has hardly been used in practice (Cousens 1987).

When is control needed? The timing of weed control is crucial in precision management. With mechanical control, the timing often determines the selectivity between crop and weeds for hoeing. Another example is found in work underway to predict the minimum lethal herbicide dosage based on the development stage of the weeds (Ketel et al 1996). On the basis of fundamental knowledge of herbicide effects at the chloroplast level (for photosynthesis inhibitors), the minimum lethal herbicide dose can be calculated for each development stage. It appears that the minimum lethal herbicide dose relates exponentially to the biomass of the weeds, indicating the strong opportunities for saving herbicide because the advised dosage is based on fairly large plants as a safety net for weed control. As a check to avoid the risk of sub-lethal dosages, a day after spraying the effect can be quantified using chlorophyll fluorescence. Further improvements can be made based on information on the effect of environmental factors on herbicide effects.

The concept of a critical period for weed control was introduced by Nieto et al (1968). This critical period represents the time interval between two separately determined components: (1) the maximum weed-infested period, or the length of time that weeds which emerge with the crop can remain uncontrolled before they begin to compete with the crop and cause yield loss, and (2) the minimum weed-free period, or the length of time that the crop must be free of weeds after sowing, in order to prevent yield losses. These two components are determined in experiments where crop yield loss is measured as a function of successive times of weed removal or emergence, respectively. The weeds in the first component are weeds that emerge more or less simultaneously with the crop, whereas the weeds in the second component are weeds that emerge later than the crop. So, basically the results of two completely different competition situations are combined.

The use of 'period thresholds' in integrated weed management systems to predict when, rather than if, weeds must be controlled to prevent yield losses was proposed by Dawson (1986). Economic period thresholds could also be calculated, indicating the length of time that a crop could tolerate weed competition before yield loss exceeded the cost of control. This would result in early-season thresholds which would denote the beginning of the critical period, and late-season thresholds denoting the end. Van Heemst (1985) demonstrated that the end of the critical period is closely related to the competitive ability of the crop. Thus, a crop with a high competitive ability has a critical period that ends early. Ecophysiological simulation models such as INTERCOM may help to analyse how such factors affect the length of the critical period.

Where should weeds be controlled? One of the main issues raised in discussions about the validity of competition models is the problem related to the assumption that the weeds are distributed homogeneously in fields. Thompson et al (1991) showed that the spatial distribution of weeds in the field has a substantial effect on the calculation of the economic threshold for weed control. Yield loss–weed density functions should be applied to the different patches separately if weeds are not homogeneously distributed. This is an obvious result of the non-linear yield loss–weed density relationship (Van Groenendael 1988). Figure 5 shows the impact of clustering of weeds on yield loss. Rice yield loss was simulated with the model INTERCOM for a range of average weed densities. The field was divided into two parts: one free of weeds and one homogeneously infested with weeds. Overall yield loss was then related to average weed density for situations in which the weed-free part encompasses 0, 30 or 60%. The clustering of weeds has a strong impact on yield loss and thus on economics and decision making.

Much is expected from site-specific weed management techniques because it is well known that the spatial pattern of annual and perennial weeds is typically aggregated (Marshall 1988, Navas & Goulard 1991, Wilson & Brain 1991, Wiles et al 1992, Cardina et al 1995, Johnson et al 1995). The spatial distribution of weeds provides a starting point for determining the perspectives of controlling only where necessary. Knowledge about this spatial distribution can be obtained by going out to the field and observing the spatial positions of the weeds. An example of the spatial pattern of individual seedlings of the weed species *Galium aparine* L. is given in Fig. 6. Weed

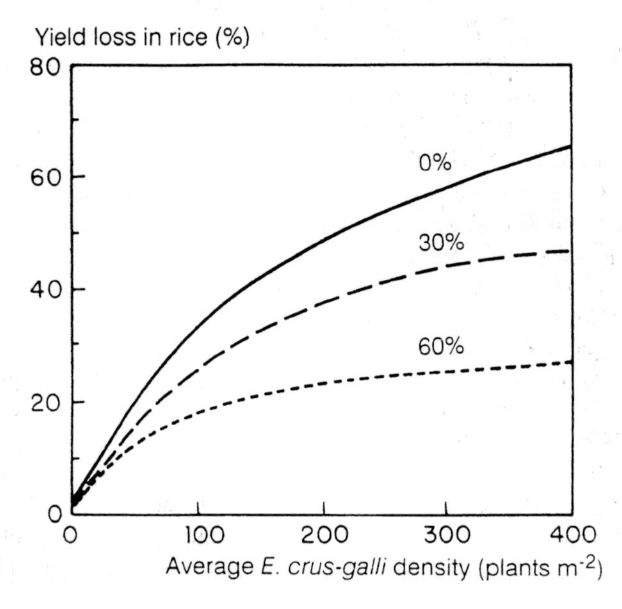

FIG. 5. Percentage yield loss as a function of average weed density when 0, 30 or 60% of the field is free of weeds and weeds are homogeneously distributed over the remainder of the field as simulated by the model INTERCOM for *E. crus-galli* in direct-seeded rice. Redrawn after Kropff & Van Laar (1993).

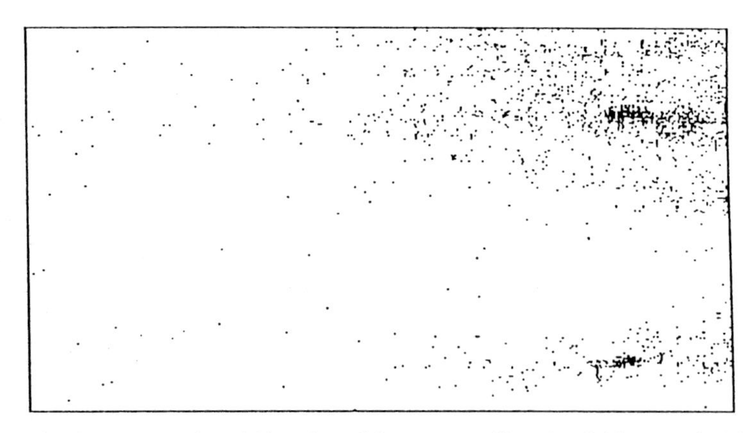

FIG. 6. An observation of spatial location of *G. aparine* seedlings in a field cropped with winter wheat (32.4 m × 18.0 m). After J. Wallinga (unpublished results).

patterns were studied on a field with loamy sand soil cropped with winter wheat, located near Wageningen, The Netherlands.

There is some evidence that patches remain stable over time (Wilson & Brain 1991, Walter 1996). It can be derived theoretically that weeds occur in patches and that patches remain stable over time using the individual-based model of annual weeds that occur endemically on an arable field with homogeneous abiotic conditions (Wallinga 1995a). This model keeps a record of the spatial position of *individual* weeds while other aspects of the weeds are as simple as in a logistic growth model. Space is conceived as a two-dimensional integer square lattice, on which individual weeds are distributed. Those individual weeds reproduce once a year. The weeds are controlled so that their density on the whole field remains low; weeds are not allowed to increase in numbers and are not eradicated. For an individual weed this means that it can be killed by weed control measures with a given probability. If it survives the weed control, it is allowed to reproduce and disperse seeds to nearby sites on the lattice. A maximum density of weeds per unit area is imposed by allowing only a limited amount of weeds per lattice site.

Starting out from randomly distributed individuals, the spatial configuration of the individual weeds rapidly settles down into a clustered pattern. This pattern is rather stable in the sense that the 'type' of clusters remains the same over time, and also in the sense that the positions of clusters show strong correlation in time, i.e. the position of clusters does not change very much over time (Wallinga 1995a). This type of behaviour shows qualitative agreement with the observation of the spatial distribution of *Alopecurus myosuroides* over a 10 year period as reported by Wilson & Brain (1991).

An implication of this modelling study is that solely on the basis of spatial population dynamics, one should expect the spatial distribution of weeds to be 'patchy' (and not uniform or random) and that these patches are 'stable' over time.

Thus one does not necessarily need external factors (like variability in soil properties) to explain the spatial distribution of weeds and stability of patches over time.

The aggregated pattern creates the potential for spraying only the weed patches, thereby reducing the amount of herbicide applied (Mortensen et al 1993). Engineering approaches have tried to develop a technology to support such a weed control (Miller et al 1995, Felton 1995). The potential reduction in herbicide use was estimated as 30% for dicotyledonous weeds and 70% for monocotyledonous weeds for maize and soybean fields in Nebraska (Johnson et al 1995), and 9–97% for cereal fields in England infested by *Elymus repens* (L.) Gould (Rew et al 1996). The use of a patch-spraying machine reduced herbicide use by 9–60% in the fallow season, and 50–80% in post-harvest application on Canadian prairies (Blackshaw 1996). The estimated reduction in herbicide use varied with weed infestation level and spatial pattern of weeds (Rew et al 1996).

The weed-free area can be used as an estimate of the potential reduction in herbicide use due to patch spraying (Johnson et al 1995). However, the estimate of the weed-free area depends crucially on the scale at which the incidence of weeds is assessed: when the field is divided into sections of 10 m × 10 m, it is likely that 0% of the area will appear free of weeds, but when each square millimetre of the field is scrutinized, over 90% of the area may appear free of weeds. There is no such a thing as 'the' weed-free area, instead, the weed-free area depends on the spatial scale of observation.

It is natural to assess the weed-free area at the spatial scale at which decisions are made, because it is only relevant to know that a square metre is free of weeds when a decision to skip herbicide application for that square metre can be made. From this point of view, patch spraying is nothing but reducing the spatial scale of decision making. The most accurate sprayers (hypothetically speaking) should spray only the leaves of weeds. The herbicide use of those very accurate patch sprayers will be a few per cent of the amount of herbicides required to spray the whole field. Patch spraying offers the opportunity, at least in principle, to decrease herbicide use to very small amounts.

It is essential to know the relation between herbicide use and accuracy of patch spraying. To obtain this relationship, the notion of scale has to be related to the spatial distribution of weeds. Therefore one needs a field where individual weeds are distributed in a particular spatial pattern. Viewing this pattern at different spatial scales is mimicked by superimposing a square grid upon the field. The mesh size of this grid (ε) determines then the spatial scale at which the field is regarded; a grid square is the smallest spatial unit for which one will make out whether it contains weeds or not. When a square does not contain a weed it is marked as 'weed-free', otherwise it is marked as 'weedy'. The perceived weedy area is the summed area of all squares that are marked as 'weedy'. Viewing the field at smaller scales is like viewing through a grid with a smaller mesh size; and with changing the mesh size (ε) the perceived weedy area changes.

As an example, such a relation between spatial scale (as mesh size ε) and weedy area is assessed for the spatial distribution of weeds as simulated by the model mentioned above. The result is plotted in Fig. 7 (note the logarithmic axes). As expected, the perceived weedy area decreases when the pattern is regarded at smaller spatial scales. Moreover, the weedy area decreases very gradually when scales become smaller. This means that there is no intrinsic spatial scale associated with the pattern, i.e. there is no typical patch size.

Since the relation depicted in Fig. 7 appears as a straight line, it can be summarized in an equation:

$$\textit{weedy area} = a \times (\textit{scale } \varepsilon)^c \tag{3}$$

where a is a constant, and the exponent c is a pattern characteristic (more specifically, it is the fractal box-counting co-dimension). For the pattern simulated by the model, c is constant for a large range of scales (Fig. 7), and the value can be estimated as $c \approx 0.9$ (Wallinga 1995b). In words, the equation says that a halving of the herbicide use requires a reduction in scale of decision making by a factor of 0.46. For example, suppose half a field is marked as weedy and requires spraying when the field is regarded in units of 10 m × 10 m, then a quarter of the field would be marked as weedy and requires spraying when the field is regarded in units of 4.6 m × 4.6 m. Reduction in herbicide use thus requires a more than proportional reduction in the scale of decision making.

The sprayed area was also calculated for the observed pattern of weeds using various mesh sizes. The relation is helpful to infer how the sprayed area reduces when buffer distance is decreased. For the observed pattern, c is approximately constant over the range of mesh sizes of 0.5–2.5 m and is estimated as $c \approx 0.5$. Halving the herbicide use requires a reduction in scale of decision making by a factor of 0.25 (J. Wallinga, unpublished results).

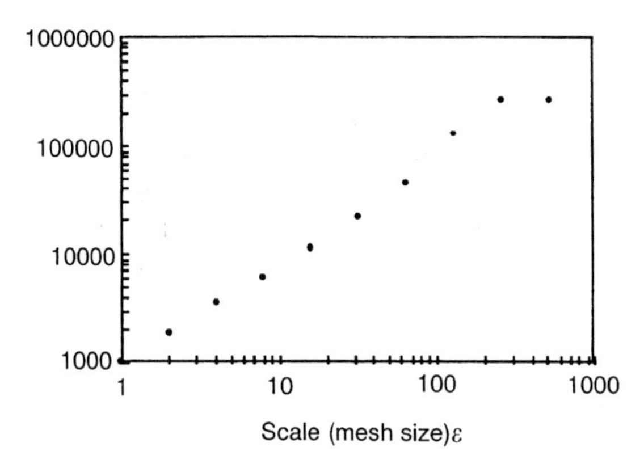

FIG. 7. The relationship between spatial scale (as mesh size ε) and the area perceived as weedy on this scale, for the simulated spatial pattern. Both weedy area and spatial scale are given in arbitrary units. After Wallinga (1995a).

The approach used here (i.e. relating scale to weedy area, as in Fig. 7) provides a method for determining the reduction in herbicide use that can be achieved when a more accurate technology would enable decision making at smaller scales.

One can go on in reducing scale of decisions and thereby reducing herbicide use up to the point where only the leaves of the weeds are sprayed. So the question 'what is the potential reduction in post-emergence herbicide application by controlling only where necessary?' results in a rather obvious answer: the fraction of herbicide use that can potentially be saved is the same as the fraction of the field that is not covered by weed leaves, in most cases very close to 100%.

Method of control. Systems that focus on herbicide selection and the optimization of herbicide dosage have been quite successful (i.e. Fischer & Lee 1981, Pandey & Medd 1991, Rydahl 1995). In general, integrated weed management systems are increasingly based on the use of mechanical (and in the future perhaps) biological control first and then using herbicides for correction.

Control

There are many ways in which control technology can be improved, ranging from precision mechanical weed management tools to precision herbicide treatments. Strategies based on the minimum lethal herbicide dose, patch-spraying equipment and new techniques for mechanical control (preferably based on optical detection) may lead to substantial reductions in herbicide use.

Conclusions

In conclusion it can be stated that options for improving weed management systems with minimum herbicide use exist in all components of the strategy: prevention, decision making and control. It is clear that the limits for such precision weed management systems are related to the technology available. Many opportunities exist for reduction of herbicide use. However, future research should not only focus on technology development, but also on prevention and strategic decision making. Quantitative insight in weed ecology and crop weed interactions is essential for that purpose.

References

Aarts HFM, De Visser CLM 1985 A management information system for weed control in winter wheat. In: Proceedings 1985 British Crop Protection Conference (BCPC): Weeds, 7A-2. BCPC, Farnham, UK, p 679–686

Auld BA, Coote BG 1980 A model of a spreading plant population. Oikos 34:287–292

Ballaré CL, Scopel AL, Ghersa CM, Sanchez RA 1987 The population ecology of *Datura ferox* in soybean crops. A simulation approach incorporating seed dispersal. Agric Ecosyst Environ 19:177–188

Barkham JP, Hance CE 1982 Population dynamics of the wild daffodil (*Narcissus pseudonarcissus*). III. Implications of a computer model of 1000 years of population change. J Ecol 70:323–344

Blackshaw RE 1996 Weed-sensing sprayer reduces herbicide use in conservation tillage. In: Brown H, Cussans GW, Devine MD et al (eds) Second International Weed Control Congress, Copenhagen. EWRS, p 1313–1316

Cardina J, Sparrow DH, McCoy EL 1995 Analysis of spatial distribution of common lambsquarters (*Chenopodium album*) in no-till soybean (*Glycine max*). Weed Sci 43:258–268

Cousens R 1985 An empirical model relating crop yield to weed and crop density and a statistical comparison with other models. J Agric Sci 105:513–521

Cousens R 1987 Theory and reality of weed control thresholds. Plant Prot Q 2:13–20

Cousens R, Mortimer AM 1995 Dynamics of weed populations. Cambridge University Press, Cambridge

Cousens R, Doyle CJ, Wilson BJ, Cussans GW 1986 Modelling the economics of controlling *Avena fatua* in winter wheat. Pest Sci 17:1–12

Cousens R, Brain P, O'Donovan JT, O'Sullivan A 1987 The use of biologically realistic equations to describe the effects of weed density and relative time of emergence of crop yield. Weed Sci. 35:720–725

Dawson JH 1986 The concept of periodic thresholds. In: Proceedings of the European Weed Research Society. Symposium on Economic Weed Control, Stuttgart-Hohenheim, Germany, p 327–331

Doyle CJ, Cousens R, Moss SR 1986 A model of the economics of controlling *Alopecurus myosuroides* Huds. in winter wheat. Crop Prot 5:143–150

Durrett R, Levin SA 1994 The importance of being discrete (and spatial). Theor Popul Biol 46:363–394

Felton WL 1995 Commercial progress in spot spraying weeds. Proceedings of the Brighton Crop Protection Conference — weeds, 1995, 8B-4. BCPC, Farnham, UK, p 1087–1096

Firbank LG, Watkinson AR 1986 Modelling the population dynamics of an arable weed and its effects upon crop yield. J Appl Ecol 23:147–159

Fischer BS, Lee RR 1981 A dynamic programming approach to the economic control of weed and disease infestations in wheat. Rev Market Agric Econ 49:175–187

González-Andújar JL, Perry JN 1995 Models for the herbicidal control of the seed bank of *Avena sterilis*: the effects of spatial and temporal heterogeneity and of dispersal. J Appl Ecol 32:578–587

Graf B, Gutierrez AP, Rakotobe O, Zahner P, Delucchi V 1990 A simulation model for the dynamics of rice growth and development. II. The competition with weeds for nitrogen and light. Agric Syst 32:367–392

Harper JL 1977 Population biology of plants. Academic Press, London

Johnson GA, Mortensen DA, Martin AR 1995 A simulation of herbicide use based on weed spatial distribution. Weed Res 35:197–205

Ketel DH, Van der Wielen M, Lotz LAP 1996 Prediction of a low dose herbicide effect from studies on binding of metribuzin to the chloroplasts of *Chenopodium album* L. Ann Appl Biol 128:519–531

Kropff MJ 1996 Strategic balancing. Inaugural address, Wageningen Agricultural University, Wageningen, The Netherlands

Kropff MJ, Spitters CJT 1991 A simple model for crop loss by weed competition on basis of early observation on relative leaf area of the weeds. Weed Res 31:97–105

Kropff MJ, Van Laar HH (eds) 1993 Modelling crop–weed interactions. CAB International, Wallingford

Kropff MJ, Vossen FJH, Spitters CJT, De Groot W 1984 Competition between a maize crop and a natural population of *Echinochloa crus-galli* L. Neth J Agric Sci 32:324–327

Kropff MJ, Weaver SE, Smits MA 1992 Use of eco-physiological models for crop–weed interference: relations amongst weed density, relative time of weed emergence, relative leaf area, and yield loss. Weed Res 40:296–301

Kropff MJ, Lotz LAP, Weaver SE, Bos HJ, Wallinga J, Migo T 1995 A two parameter model for prediction of yield loss by weed competition from early estimates of relative leaf area of the weeds. Ann Appl Biol 126:329–346

Kropff MJ, Wallinga J, Lotz LAP 1996 Weed population dynamics. In: Brown H, Cussans GW, Devine MD et al (eds) Second International Weed Control Congress, Copenhagen. EWRS, p 3–14

Lotz LAP, Groeneveld RMW, De Groot NAMA 1991 Potential for reducing herbicide inputs in sugar beet by selecting early closing cultivars. Proceedings of the Brighton Crop Protection Conference — weeds, 1991, 9A-8. BCPC, Farnham, UK, p 1241–1248

Lotz LAP, Christensen S, Cloutier D et al 1996 Prediction of the competitive effects of weeds on crop yields based on the relative leaf area of weeds. Weed Res 36:93–101

Marshall EJP 1988 Field-scale estimates of grass weed populations in arable land. Weed Res 28:191–198

Maxwell BD, Ghersa CM 1992 The influence of weed seed dispersion versus the effect of competition on crop yield. Weed Technol 6:196–204

Miller PCH, Stafford JV, Paice MER, Rew LJ 1995 The patch spraying of herbicides in arable crops. Proceedings of the Brighton Crop Protection Conference — weeds, 1995, 8B-3. BCPC, Farnham, UK, p 1077–1086

Mortensen DA, Johnson GA, Young LJ 1993 Weed distribution in agricultural fields. In: Robert P, Hurst RH (eds) Site specific crop management. ASA-CSSA-SSSA, Madison, WI, p 113–123

Navas M-L, Goulard M 1991 Spatial pattern of a clonal perennial weed, *Rubia peregrina* (Rubiaceae) in vineyards of southern France. J Appl Ecol 28:1118–1129

Niemann P 1986 Mehrjährige Anwendung des Schadensschwellen-prinzips bei der Unkrautbekämpfung auf einem landwirtschaft-lichen Betrieb. Proceedings of the European Weed Research Society Symposium 1986, Economic Weed Control, Stuttgart-Hohenheim, Germany, p 385–392

Nieto HJ, Brondo MA, Gonzalez JT 1968 Critical periods of the crop growth cycle for competition from weeds. Pest Articles & News Summ (C)14:159–166

Pacala SW, Silander JA Jr 1985 Neighborhood models of plant population dynamics. I. Single-species models of annuals. Am Nat 125:385–411

Pandey S, Medd RW 1991 A stochastic dynamic programming framework for weed control decision making: an application to *Avena fatua* L. Agric Econ 6:115–128

Rew LJ, Cussans GW 1995 Patch ecology and dynamics — how much do we know? In: Proceedings of the Brighton Crop Protection Conference — weeds, 1995, 8B-1. BCPC, Farnham, UK, p 1059–1068

Rew LJ, Cussans GW, Mugglestone MA, Miller PCH 1996 A technique for mapping the spatial distribution of *Elymus repens*, with estimates of the potential reduction in herbicide usage from patch spraying. Weed Res 36:283–292

Rydahl P 1995 Computer assisted decision making. In: Proceedings European Weed Research Society, Symposium Budapest 1995, Challenges for Weed Science in a Changing Europe, p 29–37

Schippers P, Ter Borgh SJ, Van Groenendael JM, Habekotte B 1993 What makes *Cyperus esculentus* (yellow nutsedge) an invasive species? — a spatial model approach. In: Proceedings of the Brighton Crop Protection Conference — weeds, 1995, 5A-2. BCPC, Farnham, UK, p 495–504

Silvertown J, Holtier S, Johnson J, Dale P 1992 Cellular automaton models of interspecific competition for space — the effect of pattern on process. J Ecol 80:527–534

Spitters CJT 1984 A simple simulation model for crop weed competition. 7th International Symposium on Weed Biology, Ecology and Systematics. COLUMA-EWRS, Paris, p 355–366

Spitters CJT 1989 Weeds: population dynamics, germination and competition. In: Rabbinge R, Ward SA, Van Laar HH (eds) Simulation and systems management in crop protection. Pudoc, Wageningen (Simulation Monogr) p 182–216

Spitters CJT, Aerts R 1983 Simulation of competition for light and water in crop weed associations. Aspects Appl Biol 4:467–484

Thompson BK, Weiner J, Warwick SI 1991 Size-dependent reproductive output in agricultural weeds. Can J Bot 69:442–446

Van der Weide RY, Van Groenendael JM 1990 How useful are population dynamical models: an example from *Galium aparine* L. Z Pflanzenkr Pflanzenschutz Sond XII:147–155

Van Groenendael JM 1988 Patchy distribution of weeds and some implications for modelling population dynamics. Weed Res 28:437–441

Van Heemst HDJ 1985 The influence of weed competition on crop yield. Agric Syst 18:81–93

Wahmhoff W, Heitefuss R 1988 Studies on the use of economic injury thresholds for weeds in winter barley. In: Plant research and development; a biannual collection of recent German contributions, vol 27. Institut fur Wissenschaftliche Zusammenarbeit, Tubingen, p 59–91

Wallinga J 1995a The role of space in plant population dynamics: annual weeds as an example. Oikos 74:377–383

Wallinga J 1995b A closer look at the spatial distribution of weeds—perspectives for patch spraying. In: 9th EWRS Symposium: Challenges for weed science in a changing Europe, Budapest, p 647–653

Walter AM 1996 Temporal and spatial stability of weeds. In: Brown H, Cussans GW, Devine MD et al (eds) Second International Weed Control Conference, Copenhagen. EWRS, p 125–130

Weaver SE, Smits N, Tan CS 1987 Estimating yield losses of tomato (*Lycopersicon esculentum*) caused by nightshade (*Solanum spp*) interference. Weed Sci 35:163–168

Weaver SE, Kropff MJ, Groeneveld RMW 1992 Use of eco-physiological models for crop–weed interference: the critical period of weed interference. Weed Res 40:302–307

Wiles LJ, Oliver GW, York AC, Gold HJ, Wilkerson GG 1992 Spatial distribution of broadleaf weeds in North Carolina soybean (*Glycine Max*) fields. Weed Sci 40:554–557

Wilkerson GG, Jones JW, Coble HD, Gunsolus JL 1990 SOYWEED: a simulation model of soybean and common cocklebur growth and competition. Agron J 82:1003–1010

Wilson BJ, Brain P 1991 Long-term stability of distribution of *Alopecurus myosuroides* Huds. within cereal fields. Weed Res 31:367–373

DISCUSSION

Stafford: We've generated a large number of field weed maps at Silsoe, in collaboration with colleagues at Rothamsted and the Danish Institute of Plant and Soil Science. This is necessary for our project on patch spraying. We have also developed spatiotemporal dynamic weed modelling (Day et al 1996). Are you not aware of our work?

Kropff: Of course we are aware of your work; this is referred to in our paper. What I intended to say was that the weed maps that are available are particular realizations of the kind of patterns produced by underlying ecological processes, and that it is hard to link observed patterns unambiguously to these underlying processes.

Stafford: Indeed, we have generated treatment maps to drive the patch sprayer, based on the modelling studies and field weed maps.

McBratney: Precision agriculture is really about applying information technology in agriculture. In addition to information technology, we have biotechnology at our disposal. Some 'clever' biotechnologists are producing herbicide-resistant crops which will allow us to apply herbicide uniformly to the field and not worry about where the weeds are. To me this is completely the opposite to what we should be doing with precision agriculture. As a weed scientist, what do you think about this?

Kropff: That's an important question. Two weeks ago we had a meeting of weed scientists from the EC to discuss this issue. It seems to some that a herbicide-resistant crop would solve all our problems: you spray everything with glyphosate (a broad-spectrum herbicide) and all the weeds are killed. One problem with this approach is that not all weeds are controlled by glyphosate; another is that herbicide resistance is starting to appear. *Lolium rigidum* in Australia is the first weed to become resistant to glyphosate. So I don't think that the use of herbicide-resistant crops is going to be a very sustainable solution. In some situations it may help as one method of weed management, but it's not the way forward. We need to base future weed management systems on agroecological knowledge to prevent weed problems in the first place.

McBratney: Philosophically I still have trouble with the whole concept, because it seems to be an imprecise method of management.

Groffman: It sells many gallons of glyphosate.

Rabbinge: This approach is not necessarily in conflict with precision agriculture, because you can fine-tune your measures not in space but in time. In this way it is possible to reduce pesticide use considerably, which is a widely accepted goal.

Kropff: And from an environmental point of view it's more beneficial because glyphosate is less damaging than some of the other herbicides currently used. However, preventive options should not be neglected.

Robert: I have been told by colleagues working on biotechnology that genetically modified seeds shouldn't be used for entire parcels but instead islands of conventional seeds should also be used to reduce the risk of pests developing resistance. Precision agriculture may also play a role in selecting zones of bioengineered seeds and conventional seeds.

Stafford: I would like to try to turn around the gene manipulation issue to relate it to precision management. One area that we've been thinking about, which relates to your comment about the difficulty of detecting weeds, is whether or not we could put a marker in the crop to enhance the contrast between the crop and other vegetative material (namely weeds) in some part of the UV/visible/IR spectrum. This would ease the weed detection problem.

Kropff: One option could be the use of crops with non-green pigments that facilitate photosynthesis as well. We used this concept in our competition studies in rice in which we used purple-coloured rice varieties as a weed to make effects visible. The purple rice grows just as well as normal crops. If this could be introduced to distinguish the crop from the weeds on the basis of optical properties, we don't even need biotechnology to address this problem.

van Meirvenne: In this respect, image analysis shows promise. About a year ago I heard a talk by Mats Rudemo from Copenhagen who was doing research on identifying weeds from video images. The proposed application of this is for the farmer to drive over the fields with a video camera half a metre above the soil, analyse the image, identify and localize the weeds and then spray directly onto the weeds alone.

Robert: Different kinds of weed sensors are in development, but two systems have already been commercialized in the USA. One type, based on real time image analysis, detects weeds between plant rows by means of cameras attached to the sprayer boom. The other type, based on the intersection of a light beam, detects weeds taller than the plant canopy. In both cases, the detection of weeds is followed by the local application of herbicide.

Kropff: Indeed, there is a lot of research going on in terms of control based on observation with sensors. In Wageningen, agrotechnologists have developed equipment that detects whether a plant is a crop plant or a weed plant. If it is a weed this equipment has a mechanical way of controlling it. The machine doesn't need to recognize what kind of weed it is. The key is that the crop has to be planted on fixed grids, or that the crop can be detected and distinguished from the weeds.

Goense: I can give some background on this technique, which was developed in our department. It incorporates a light beam and we make use of the fact that many crops are precision planted. If, for example, we plant sugar beets 20 cm apart, we know the distribution. When you analyse the pattern of the breaks in the light beam, you recognize exactly where there should be a sugar beet plant and you can get rid of all the others.

In agricultural engineering we are currently facing a new problem. Five years ago, computing power was the limiting factor in this technology, whereas now it isn't: the limiting factor is instead the mechanics, getting the tools to move fast enough to cut the weeds between the sugar beets and to stop in time to avoid chopping down the crop plants. One of our students developed the elegant method of using a rotating wheel with two knives pulled in by springs. Just by changing the rotational speeds a fraction, the forces are enough to protract or retract the knives.

Groffman: We've heard a bit about ecological or 'organic' farmers. They don't use herbicides. There are things that precision agriculture can offer to these ecological farmers. I think it's important to articulate them because these people are very imaginative and they often have interesting ideas. The best example is conservation tillage, which from a sustainability point of view and a environmental point of view is a wonderful thing, but the ecological farmers won't do it because it requires the use of herbicides. It's useful to articulate what precision agriculture has to offer to people outside the mainstream of agriculture, to bring those people in.

Rabbinge: In The Netherlands we have developed mixed farming systems in which we are trying to develop both technological and ecological methods. There is cross-fertilization between the ecologists and the ecotechnologists. It has resulted in a substantial increase in efficiency and a reduction in negative environmental side-effects.

Kropff: The ecological farmers we are working with in one of our research programs have already developed many ways of dealing with weeds. Of course, they have had to, because this has been their main problem (hours needed for hand weeding). They have also developed mechanical control methods. However, they still mention weeds as their primary problem.

Stafford: The advances in technology raise a philosophical question with regards to spraying. You suggested that real time sensing and spraying could be the order of the day. Several years ago we attempted to do that and came up against the processing problem: we just could not process the information fast enough. Since then, we have realized that this might not be the right approach: a rather better approach is to get a 'bird's eye' view of what is happening in the field with regard to the weed patch dynamics. If you just sense what is immediately ahead of the tractor and spray as a result of that, only a very simple herbicide application strategy can be implemented. It is preferable to know what the population distribution is within the field — in other words, to develop a historical weed map — and to develop the herbicide strategy on the basis of that and then spray (Stafford & Miller 1993, Miller et al 1995). As you said, weed patches tend to remain in the same location from year to year, so the same weed map can be used over several seasons and modified as more weed information becomes available.

Rabbinge: This is support for the combination of strategic, tactical and operational decision making.

Webster: I would like to comment on the question of detecting weeds, whether you do it by image analysis or whether you do it by people. It's great when the weeds you want to kill are substantially different from the crop plant. But when it comes to detecting something like black grass in wheat, or *Echinochloa* in rice, you're in trouble. The only way that they have been able to do this at Rothamsted is to build a device where people sit on a platform mounted on a vehicle and record the densities of the weeds as the vehicle moves slowly through the crop (Rew et al 1996).

Kropff: You are right. Therefore you need strategies such as the use of pigmented crop plants in the case of look-alikes. The purple rice varieties I mentioned earlier are now being used now in Spain and Italy, because a common weed is a wild so-called 'red rice' variety that loses its seeds very quickly. They're now using these purple varieties in rotation, and everything that is not purple is removed. I think this type of approach is very important in the case of weeds that look exactly like the crop.

Robert: I should add that a colleague working on biotechnology mentioned recently that new ways of differentiating plants from weeds could be available in the not-too-distant future. For example, it will be possible that through plant gene manipulation crop plants will emit a light signal of a selected wavelength. Sensors installed on a herbicide spreader will differentiate crop plants emitting a 'good' signal versus weeds, and direct application accordingly. This could also be developed for other kinds of pests.

References

Day W, Paice MER, Audsley A 1996 Modelling weed control under spatially selective spraying. Acta Horticulturae 406:281–288

Rew LJ, Cussans GW, Mugglestone MA, Miller PCH 1996 A technique for mapping the spatial distribution of *Elymus repens*, with estimates of the potential reduction of herbicidal usage from patch spraying. Weed Res 36:283–292

Miller PCH, Stafford JV, Paice MER, Rew LJ 1995 The patch spraying of herbicides in arable crops. In: Proceedings Brighton Crop Protection Conference, November 1995. British Crop Protection Council 3:1077–1086

Stafford JV, Miller PCH 1993 Spatially selective application of herbicide to cereal crops. Comput Electron Agric 9:217–229

General discussion III

Rabbinge: We began today with a provocative paper by Vic Barnett on the performance of wheat models. Since then we have seen examples where models may be of use, although not for predictive purposes. We are defining more clearly the conditions that must be fulfilled for a model to be of value. Otherwise, we end up with modelling for the sake of it. The other subject we discussed extensively concerned examples where precision agriculture can be of use. There are some questions still outstanding from this. Finally, we have had contributions from geostatistics and spatiotemporal statistics. How are we to combine the different tools — GIS, sampling techniques and the statistics behind these — in order to come up with systems that are practical and will be operational within the next five years or so?

Bouma: One area where I think we would all agree that modelling has been successful is soil water content. We can simulate soil moisture regimes pretty well. We find that the water regime as a function of climatic and boundary conditions is related to practically everything we are concerned about: crop growth, nitrogen transformation, pests, diseases, etc. It is a central theme from which other things are derived. It is also one of the variables susceptible to management.

Burrough: Are you, from a position of soil water modelling, able to specify the kinds of spatial and temporal relationships that are necessary to support that modelling, in such a way that one could choose the appropriate sampling techniques, geostatistical techniques and special data analysis techniques to reach your purposes? That would be a different approach from going down the computer store to buy a proprietary GIS and saying 'Sorry, this is the way we've got to work because that's what it does for us'. Are you able to specify the mathematical operations you need in such a generic way that those can be translated? If you can do that then we can go a long way.

Bouma: Yes. Various publications document this.

Mulla: I wanted to comment on Alex McBratney's paper. My question relates to his null hypothesis, which may have left us with somewhat of a negative outlook. I'm very interested in knowing a slightly different answer to the question on temporal variability: that is, are the spatial patterns in yield for similar hydrological regimes stable over time? I'm not that interested in knowing whether or not the spatial patterns are stable from year to year, because perhaps as time passes we may develop our ability do some very general weather prediction in the sense of whether the year ahead is going to be excessively wet or dry. If we know that the upcoming year is going to have a certain general characteristic, are the spatial patterns in yield for years that have that general characteristic going to be stable?

McBratney: One would imagine so. My final statement wasn't meant to be a negative one, it was just meant to make us, as scientists, take stock of where we're at. We don't want the 'Emperor's new clothes' syndrome applying to precision agriculture. I do think there's something in it, but clearly if you were to have a field that was spatially uniform every year, you would not want to manage it differentially. In the same way, if through time you can't do any kind of prediction, you need to apply some kind of uniform management. You're playing a game of risk, and your risk-aversion strategy says that if you don't know what's going to happen, you manage uniformly.

This may be controversial, but I think the science of precision agriculture is not good at the moment. It's important that we demonstrate and document that variable management can do better than uniform management in many situations. We have got to do that because there are just not enough of those data around.

Voltz: From the point of view of risk management it is not always the best to try to get uniform and optimal development of the crop over a field. An example of this is in non-irrigated farming systems where water resources are uncertain. For some species, it is known that drought affects plants with large leaf area more severely than those with small leaf area because the water needs of the former are larger. So if a farmer succeeds in obtaining uniform maximal crop development in the early part of the cropping season, and later on a drought occurs, he risks losing his entire crop. In contrast, if there is some heterogeneity in crop development the drought may affect severely only part of the crop. In terms of strategy and minimizing risk (farmers are often more interested in minimizing risk than optimizing everything) the latter strategy may be better because you don't know in advance what the weather will be.

Webster: I should like to pursue this theme of risk. All of us in this room are still thinking of western 'hi-tech' agriculture. Small farmers in Africa, for example, display quite purposeful non-uniform treatment, and that is also to minimize risk. Another example is in the wet–dry tropics where the farmer deliberately leaves some of his land bare so the water runs off onto other patches where the crop is allowed to grow. Here, his strategy is again minimizing the risk, because if that water infiltrates uniformly everywhere he might get nothing.

Rabbinge: Many of the traditional agriculture systems in for example Western Sub-Saharan Africa have been developed over many centuries in order to minimize risk. But at the same time you have to conclude that due to the much heavier productivity demand it's impossible to maintain these systems in the traditional way where the output per unit of input is too low to satisfy the demand. A technology jump is needed to attain higher productivity levels and this requires better insight and fine-tuning of the ecological potential to the new demands. This is happening dramatically in rice cultivation at the moment, with the move from irrigated rice to upland rice. A hectare of rice requires between 600–800 h labour; within the next 10 years this will be reduced to a maximum of 60 h. This dramatic change requires appropriate technology and management.

Webster: Rice is a very special crop; I was thinking more of millet, in particular, in dry climates.

Rabbinge: Millet is a typical example of a crop for very marginal areas. This year ICRISAT received a prize for the improvement of millet cultivation. They developed fine-tuned systems for millet under harsh conditions and fulfilled a demand of the farmers, the end users, for an appropriate risk-averse management system.

Meshalkina: There are two special problems related to the development of precision agriculture. The first concerns how we should summarize the initial information about a field that the farmer already has. The second involves the strategic goal of precision management technology in a given case. Is it to reduce the heterogeneity in a field or to maintain the differences that exist between areas in order to use them effectively?

Rabbinge: In the past heterogeneity in space and time was seen as a liability, and we tried to get rid of it immediately. Many agricultural practices were designed to bring about uniformity. Nowadays, because of new insights, equipment and technologies, it is possible for us to use that heterogeneity, so it can now be seen as an asset.

Meshalkina: Usually we are thinking in terms of different spatial scales. For example, if the minimal contour in a field is about 20 m, we probably need to reduce the variability in smaller distances.

Rabbinge: Over the last couple of days we have been considering the heterogeneity at different aggregation levels: on the field level you are talking about a scale of metres. On a regional level the scale is in kilometres and on a global level it involves grids of 10 km^2. This requires other techniques and approaches. During this meeting we have concentrated purely on the individual field levels and not the higher aggregation levels. However, the higher levels are very appealing for the future and I hope that they will also be considered soon.

Simmelsgaard: Earlier we discussed stability in yields from year to year within a field. We have done a correlation of yield maps from two 10 ha fields in three years, and we found an r^2 from 10–30%. One student has analysed the yield maps from 82 fields from Denmark and UK, and he found r^2 values varying from 0–81%. For years after each other, the mean r^2 was 41% for cereals. For one year or more between the same crop and between different crops the correlation was smaller.

Rabbinge: If you consider the longer term and not just a two year sequence, nearly everywhere in the world there has been a productivity rise over the last 20–50 years. I don't know how much longer this can continue. In Western Europe have we reached very high yield levels and a further jump requires new crop plants and different management methods. The majority of agriculture in the world still has yields far below what can potentially be achieved with modern technologies and varieties.

In the late 1960s when we did calculations on maximum potential yield we came up with estimations of wheat yields in the order of 10 tonnes/ha. At the time yields were about half this and everyone considered this potential level to be ridiculously high. Twenty-five years later the average yield in The Netherlands is already 9 tonnes/ha. Had we not done those calculations no one would have expected such yields to be achievable. The reason this level was not reached earlier was mainly because of inappropriate pest and disease control and agronomic measures.

Optimal mapping of site-specific multivariate soil properties

Peter A. Burrough and Julian Swindell*

*Netherlands Centre for Geoecology (ICG), Department of Physical Geography, Faculty of Geographical Sciences, PO Box 80115, 3508 TC Utrecht, The Netherlands and *Royal Agricultural College, Cirencester, Gloucestershire GL7 6JS, UK*

Abstract. This paper demonstrates how geostatistics and fuzzy k-means classification can be used together to improve our practical understanding of crop yield-site response. Two aspects of soil are important for precision farming: (a) sensible classes for a given crop, and (b) their spatial variation. Local site classifications are more sensitive than general taxonomies and can be provided by the method of fuzzy k-means to transform a multivariate data set with i attributes measured at n sites into k overlapping classes; each site has a membership value m_k for each class in the range 0–1. Soil variation is of interest when conditions vary over patches manageable by agricultural machinery. The spatial variation of each of the k classes can be analysed by computing the variograms of m_k over the n sites. Memberships for each of the k classes can be mapped by ordinary kriging. Areas of class dominance and the transition zones between them can be identified by an inter-class confusion index; reducing the zones to boundaries gives crisp maps of dominant soil groups that can be used to guide precision farming equipment. Automation of the procedure is straightforward given sufficient data. Time variations in soil properties can be automatically incorporated in the computation of membership values. The procedures are illustrated with multi-year crop yield data collected from a 5 ha demonstration field at the Royal Agricultural College in Cirencester, UK.

1997 Precision agriculture: spatial and temporal variability of environmental quality. Wiley, Chichester (Ciba Foundation Symposium 210) p 208–220

The aim of precision farming research is to help farmers optimize the management of their fields so that gross returns are maximized for each management area. To do this, the farmer needs to know how the conditions for growing crops vary over the area of interest. Research into the modelling of crop response to site conditions has produced many mathematical models (Dumanski & Onofrei 1989) but, as Varcoe (1990) has pointed out, citing Fischer (1985), there is often a credibility gap between modelling and practical decision making. Another problem is that computer simulation models require large amounts of data that can often only be collected routinely at experimental farms, or by commissioning extra surveys. A third and very often serious limitation of crop yield models is that they use data that necessarily have been collected for small areas or volumes of soil that can reasonably be sampled for laboratory analysis. If the spatial variation of soil is considerable (and there is much evidence to show that soil properties vary spatially and temporally at all levels of resolution from millimetres to hundreds of kilometres — Burrough 1993) then the model predictions may also vary spatially in ways that cause difficulties for practical users.

When using computer modelling to predict yield variations over an area we can adopt several different strategies. We can collect the basic data to run the crop yield models from a set of representative sample sites and then we can (a) model the yields first and then interpolate them over the area, or (b) we can interpolate the basic data first and then compute the model over the whole area. Both options are possible using geographical information systems (GIS), but both require considerable amounts of extra data and the procedures involved are both theoretically and practically difficult for a farmer. Nobody wants to pay for resampling of fields for soil properties and chemical tests, nor for airborne remote sensing of crops for water stress and disease, unless strictly necessary.

Therefore, computer modelling of predicted crop response to site conditions is good science but difficult to turn into a practical tool. Even if the modelling and GIS operations can be presented to the farmer as hard-wired, push-button modules, there is still the problem of collecting the data, because that is an extra cost, which is usually quite considerable. On the other hand, the farmer has to go over his field several times a season to plough, sow, fertilize, spray for weeds and pests, and to harvest. It is technically quite possible to fit his tractor with automatic sensors that record the location and attributes of the crop during routine operations to provide data that can be useful for management. This is done by using locally referenced GPS (global positioning systems; Kennedy 1996) to determine the absolute location and electronic data recording techniques to determine crop attributes, all of which are written on a smart-card for post-operation plotting and analysis. The methods are now operational, though still requiring improvement (Swindell 1995).

Automating the sampling for the farmer by linking it to an operation he has to do anyway minimizes his extra costs to the GPS system and the post-processing analysis: these are small compared with data collection and may increase profits through improved ability to manage the crop well. The simplest example of linking GPS technology to agricultural operations is the automatic recording of crop yields, and these data we use here. With a little thought it should be possible to record other attributes of crop growth (colour, height, disease, etc.) so that the kinds of spatial analysis presented here can be done during the whole of the growing season.

Thus we can summarize the aim of this paper as:

> The creation of simple, but informative maps of crop yield-site response from historical, detailed, yet easy to collect data so that farmers can optimize crop management over a field.

For reasons of cost-effectiveness and spatial resolution it is essential that large amounts of data can be collected quickly and cheaply. In this case the data are obtained from instruments placed on farm equipment such as tractors and combine harvesters linked to GPS location equipment in order to record initial data on seasonal or multi-year and multicrop spatial patterns. For the farmer, the reasons for adopting this approach are:

(1) The use of GPS linked to sensors on standard farm equipment is an easy way to collect huge amounts of data that directly reflect the integrated crop response to the conditions in the field.
(2) The farmer is not interested in crop yield variation over very short distances, and certainly not at those distances that are of similar dimensions to the size of supports used for soil sampling because these fall within the size of his machinery (up to 5 m × 5 m). The farmer is also not interested in variation over the landscape at large (>1000 m). If he is interested at all, then it is in the variations in crop response/yield within his main spatial units (i.e. fields).

An alternative approach would be to use crop models to predict crop response. We believe that for detailed field management this is also not a suitable option (cf. Varcoe 1990). First, models are usually point predictors — they may vary in complexity from logical to empirical to deterministic based on understanding of growth processes — and they are usually fed with point data. Soil data etc. are usually measured on a support that is very much smaller than the minimum area (5 m × 5 m) that a farmer can deal with. If there is significant soil variation then there will be significant spatial variation in the yield predicted at point locations. To capture this spatial variation requires large numbers of points to be modelled.

Second, crop yield is a non-linear response to environmental conditions — a cold wet spring can retard crop development on poorly drained sites but these can support better growth during a dry summer. Actual crop yield measurements and other field-based observations integrate over space and time.

Third, it is useful to have a method that can easily incorporate other measurements during the growing season, such as water stress, nutrient deficiency or incidence of disease.

Theory

The methods proposed in this study use both the principles of geostatistical interpolation (kriging) and continuous classification (fuzzy *k*-means). Geostatistical methods are used to bring all attribute data to the same coordinate system, in this case a regular grid. The method of fuzzy *k*-means (McBratney & de Gruijter 1992, Vriend & van Gaans 1994) is used to compute a limited set of overlapping multivariate classes from the attribute data. This yields a set of overlapping continuous classes in which a given site has a continuous membership value ranging from 0 (definitely out) to 1 (definitely in).

The Membership m of the ith object to the cth cluster in ordinary fuzzy k-means, with d the distance measure used for similarity, and the fuzzy exponent q determining the amount of fuzziness, can be computed using Picard iteration of Bezdek's equations (Bezdek 1981):

$$m_{ic} = \left[d_{ic}^{-2/(q-1)}\right] \Big/ \left[\sum_{c=1}^{k} d_{ic}^{-2/(q-1)}\right] \quad i = 1, \ldots, n;\ c = 1, \ldots, k \tag{1}$$

and

$$\mathbf{c}_c = \left[\sum_{i=1}^{n} m_{ic}^{q} \mathbf{x}_i\right] \Big/ \left[\sum_{i=1}^{n} m_{ic}^{q}\right] \quad c = 1, \ldots, k \tag{2}$$

where $\mathbf{x}_i=(x_{i1}, \ldots, x_{ip})^{\mathrm{T}}$ is the data vector for individual i, and $\mathbf{c}_c=(c_{c1}, \ldots, c_{cp})^{\mathrm{T}}$ is the vector representing the centre of class c, and $d^2(\mathbf{x}_i, \mathbf{c}_c)$ is the square distance between $\mathbf{x}_i$ and $\mathbf{c}_c$ according to a chosen definition of distance.

The user must:

- Choose a number of classes k, with $1<k<n$ (where n is the total number of observations).
- Choose a value for the fuzziness exponent q (usually $q=1.5$ is sufficient).
- Choose a definition of distance in the variable space (Euclidean, Mahalonobis, etc).
- Choose a value for convergence.
- Initialize the clusters with random memberships of central values from a hard k-means partition (McBratney & de Gruijter 1992).

This procedure yields a fuzzy membership value for each data point and for each class, subject to the condition that $\Sigma(m_{ic})=1$. These continuous membership values can be interpolated by ordinary kriging to a fine grid to map the possibility of belonging to each class (see de Gruijter et al [1997] for interpolation methods that constrain joint interpolation to $\Sigma(m_{ic})=1$).

One of the main problems with interpolating fuzzy k-means is that there are as many maps as classes. However, there is a very simple procedure for computing the dominant class for any given site, either as a continuous or a choropleth map. The boundaries of the different classes can be extracted automatically from the intersections of the continuous surfaces for each class using a method proposed by Burrough et al (1997). The method uses the simple concept that at sites having near equal membership values for more than one class, confusion about which class to allocate the point is greatest. If these points of greatest confusion are linked they naturally define the boundaries between the zones dominated by any given class. The confusion index is computed from the interpolated maps using:

$$CF_{xi} = (1 - (\max(m_{j,xi}) - \max(m_{k,xi})))\tag{3}$$

where $\max(m_{j,xi})$ is the maximum value of j classes at site x_i and $\max(m_{k,xi})$ is the next maximum value of the remaining k classes at the same site. The CF_{xi} value ranges from 0 (no confusion — one class is completely dominant) to 1 (two or more classes have the same maximum membership value and allocation is impossible).

If the original surfaces for each membership function vary smoothly over space then the map of the confusion index has well-defined, narrow zones or ridges where the membership value surfaces for different classes intersect. These ridges can easily be converted to hard boundaries by the use of automated image analysis methods. The confusion index map is therefore a simple, useful and important instrument for the segmentation of the multivariate space of interest, something that has only previously been achieved for one-dimensional transects (Wackernagel et al 1988).

Once boundaries have been found and extracted, the map of yield variation can easily be simplified to a choropleth map that provides the farmer with the information needed.

The study area

The study area is one of the 5 ha demonstration plots on Field 15 (Driffield Bank) at Harnhill Manor Farm in Gloucester, UK. This 30 ha field is the site of a long-term integrated cropping systems experiment run by the Royal Agricultural College in Cirencester, together with the AFRC Long Ashton Research Station, which is part of

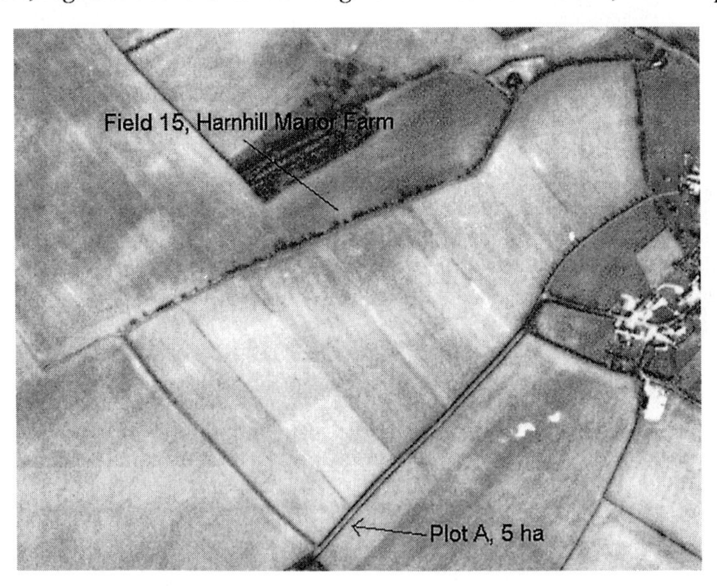

FIG. 1. Location of the 5 ha field at Harnhill Manor Farm, Gloucester, UK.

a larger European Union-funded study (Davies & Limb 1995). The field is divided into six demonstration plots of equal area, the first of which provided the data for this study (Fig. 1). The trials are managed using low-input techniques, which include special cultivation systems, low agrochemical inputs, crop–fallow rotations and the encouragement of beneficial arthropods. Field 15 is fairly level, with a gentle slope towards its north western side. It is underlain by Forest Marble Clay and Cornbrash Oolitic Limestone; mapped soils include the Sherborne, Moreton, Evesham, Haselor series and an unclassified alluvial soil. Sherborne and Moreton are similar soils, with Sherborne being very shallow over limestone. Evesham and Haselor are stone-free, clay textured and poorly structured: Haselor series overlays limestone at shallow depths (Conway 1995).

A digital elevation model (DEM) (Fig. 2a) was made by digitizing contours on a 1:10 000 topographic map and interpolating the levels to a 2.5 m grid covering the whole field.

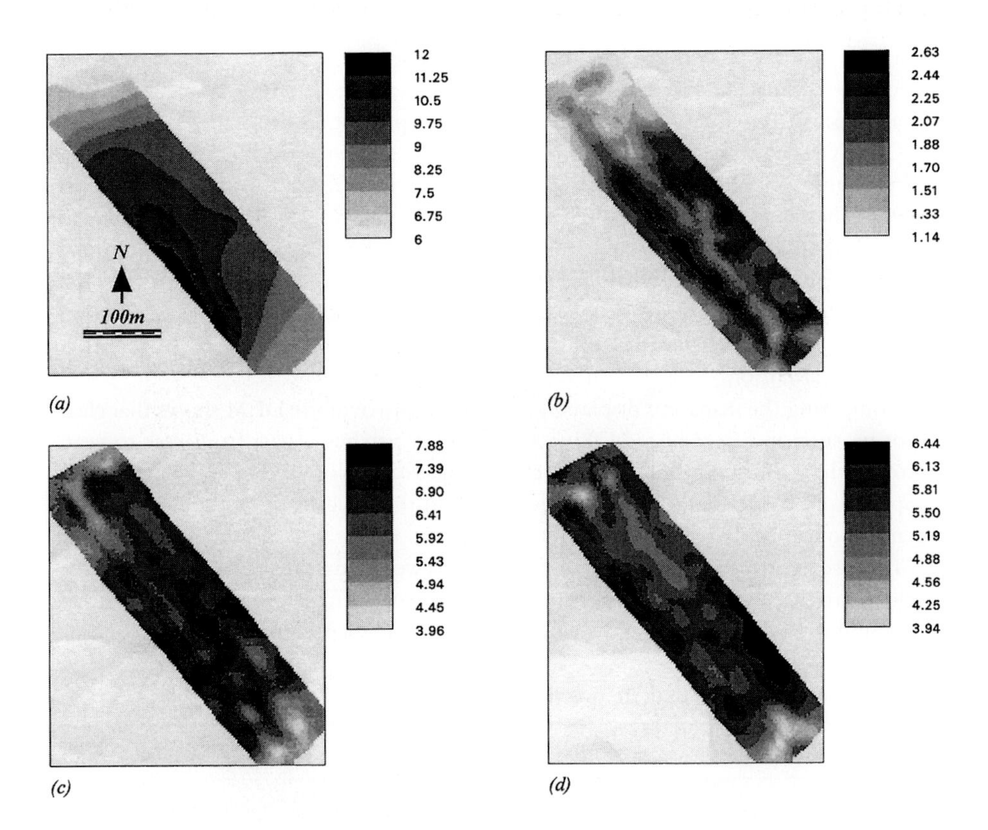

FIG. 2. Interpolated yields. (a) Digital elevation model (scale in metres). (b–d) Yield variation maps (scale in tonnes/ha) for (b) rapeseed, 1993; (c) winter wheat, 1994; and (d) winter barley, 1995.

Data collection

Yield data were collected for three years and for three different crops (1993 oilseed rape; 1994 winter wheat; 1995 winter barley). Crop yield data were collected using a GPS-located combine harvester referenced to a local base station located at the Royal Agricultural College in Cirencester, some 7 km away (Swindell 1995). Yield data were recorded on a smart card every 1.2 sec as the combine harvested the field, which is equivalent to a yield measurement every 2.5 m; this gave a database of some 1700–2500 points for each year with a support measuring 2.5 m long and 5 m wide. Due to errors and instability in GPS measurements, locations are often in error and had to be adjusted to the common coordinate base used for the digital elevation model. Yields can also be underestimated if an area is only partially harvested or harvested twice. Table 1 gives summary statistics of the raw data.

Data processing

For each year it was necessary to smooth out the irregularities in the data collection and bring everything to a common base. This was done by converting the original data from smart card to base XYZ coordinates using a spreadsheet program.

Crop yields for each year were interpolated to a 2.5 m grid by first computing variograms and fitting appropriate models using the GSTAT program (Pebesma 1996). The raw yield data were interpolated as averages for 20 m blocks in order to remove the worst effects of short-range variation due to harvesting and locational errors. Figures 2 b–d show the resulting generalized yield variation maps for each year. Note that if the crop yield variograms had showed poor spatial correlation structures (large nugget variance) one might decide that there is no spatial variation worthy of attention.

Comparing the maps and displaying them as drapes over the DEM shows that clear features such as stripes of reduced yield and other patches persist from year to year, though there are strong differences between the years. Not all, but some differences appear to be associated with local elevation, and the lower ends of the field have in general lower yields.

Because these maps have large numbers of cells we subsampled them using a regular grid to provide a training set of points for the multivariate fuzzy *k*-means classification. A sampling grid was laid over the yield maps to collect the interpolated yields for each

TABLE 1 Basic yield data (tonnes/ha)

Crop	*n*	*Min*	*Max*	*Mean*	*SEM*	*SD*
Rape 1993	1989	1.00	2.80	1.985	0.0099	0.44
Winter wheat 1994	2199	3.00	8.80	6.442	0.0238	1.12
Winter barley 1995	1700	2.60	7.60	5.485	0.0182	0.75

year at some 84 locations and read them into a single data file. Scatterplots were drawn for the yields for each pair of years (Fig. 3) and the data suggested that the field had responded differently for grains than for oilseed rape. These sampled data were then classified into four classes using the FUZNLM program (Vriend & van Gaans 1994). The yield data for each pair of years were plotted (1993/1994, 1993/1995, 1994/1995) using the hard class indicator from the fuzzy *k*-means to label the points on the scatter plot (Fig. 3). We chose four classes because these can easily be interpreted in simple terms such as 'good for all crops', 'low for all crops', 'good for cereals, average–poor for oilseed rape', and 'good for oilseed rape, average–poor for wheat'. Table 2 gives an overview of these classes. The Scheffé *post hoc* probabilities indicate which pairs of classes for each harvest most overlap.

Variograms of membership values were calculated for each fuzzy class using the data from the training set and then each class was interpolated to the 2.5 m grid. Figure 4 shows the results; note that the spatial patterns vary quite coherently. Using standard GIS raster operations (Burrough & McDonnell 1997, Wesseling et al 1996) the

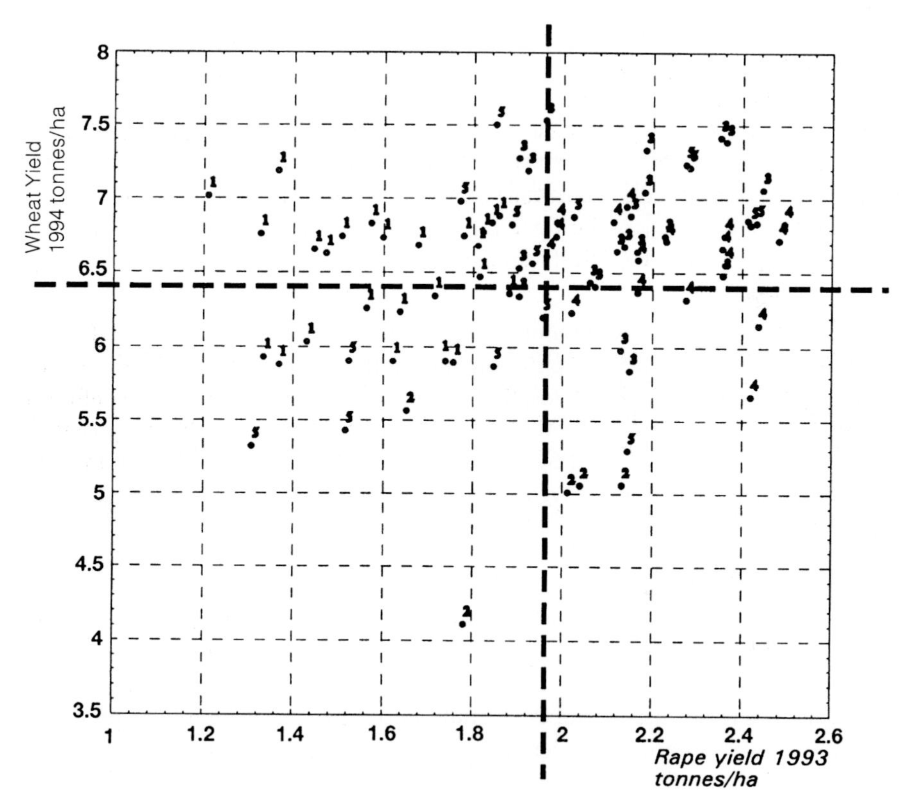

FIG. 3. Scattergram of yield with class numbers. 5 represents an intergrade. Dashed lines indicate means.

TABLE 2 Statistics of fuzzy classes

Crop/year	*Class/n*	*Mean yield*	*SD*
Rape 1993	1–24	1.60	0.20
	2–5	1.92	0.20
	3–20	2.13	0.16
	4–18	2.24	0.18
	Intergrades–16	1.93	0.33
Wheat 1994	1–24	6.52	0.38
	2–5	4.96	0.53
	3–20	6.79	0.48
	4–18	6.56	0.32
	Intergrades–16	6.42	0.72
Barley 1994	1–24	5.42	0.28
	2–5	4.66	0.50
	3–20	6.02	0.17
	4–18	5.40	0.18
	Intergrades–16	5.54	0.23

Scheffé post-hoc probabilities of class overlap (5 indicates intergrades):

Rape 1993:	classes 2–3, 0.45;	2–5, 0.999;	3–4, 0.676.	
Wheat 1994:	classes 1–3, 0.497;	1–4, 0.997;	1–5, 0.984;	3–4, 0.767.
Barley 1995:	classes 1–4, 0.999;	1–5, 0.626;	4–5, 0.576.	

maximum fuzzy class values and confusion index were computed over the field (Fig. 5). The second derivative of the confusion surface (profile curvature—Burrough & McDonnell 1997) was used to locate the maximum rates of change of confusion index and thereby to determine the optimal boundaries between the fuzzy classes. The boundaries were extracted by simple logical operations and the fuzzy surfaces were converted to hard classes. Figure 6 shows the results.

Discussion and conclusions

The composite maps of both continuous membership values and hard classes clearly distinguish the different areas of the field and summarize the crop response. These maps can be used easily and directly to adjust management practices if required. Because the whole analysis is data driven and can be fully automated, new maps can be produced any time new data are acquired. These data could be crop yields in future years, but they could also be data collected from other sensors mounted on tractors to monitor crop colour, temperature, moisture stress, disease or other problems that occur during the season. If these kind of data were used then the result of the analysis could be used to

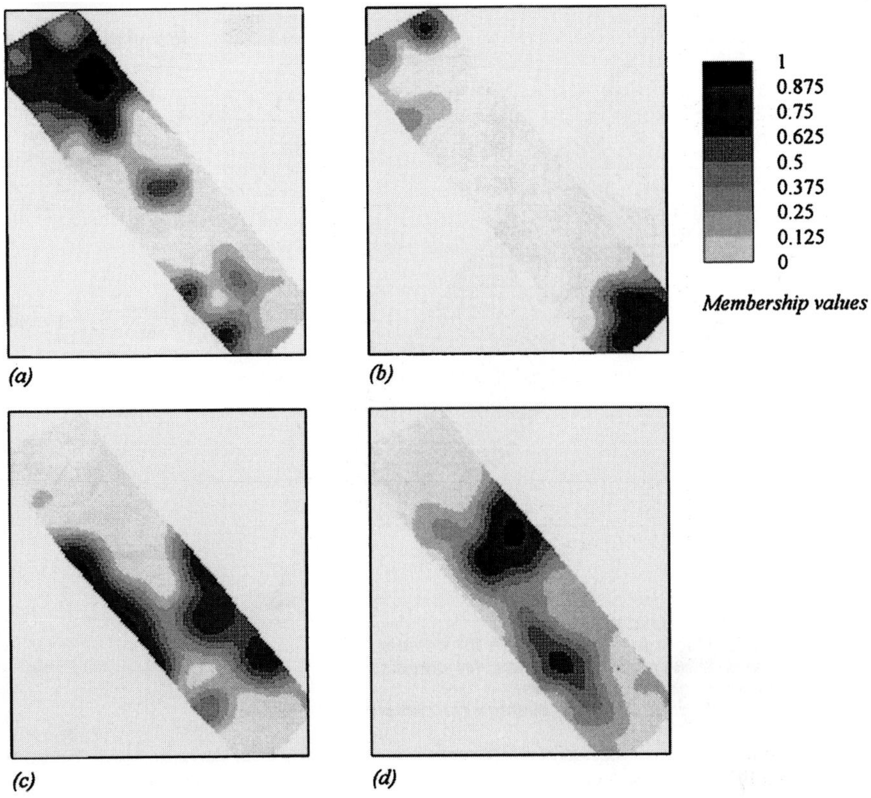

FIG. 4. Maps of membership values for each fuzzy class interpolated to the 2.5 m grid. (a) Fuzzy class 1; (b) class 2; (c) class 3; and (d) class 4.

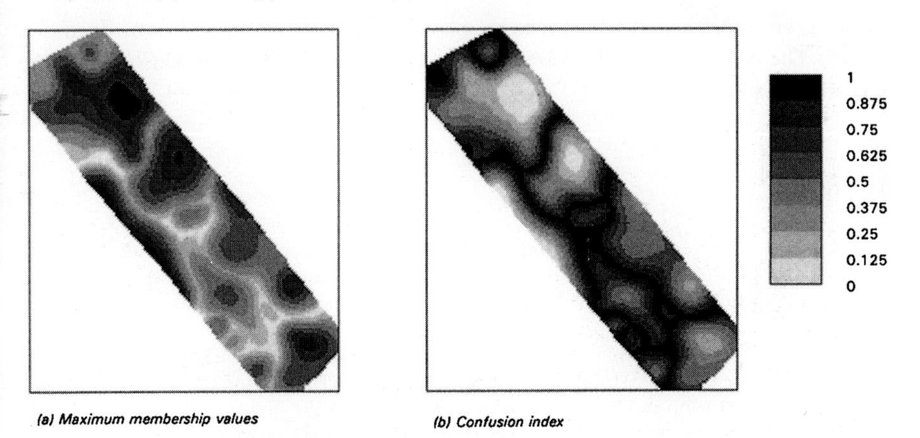

FIG. 5. (a) Maximum fuzzy class membership values, and (b) confusion index, calculated for the field using standard GIS raster operations.

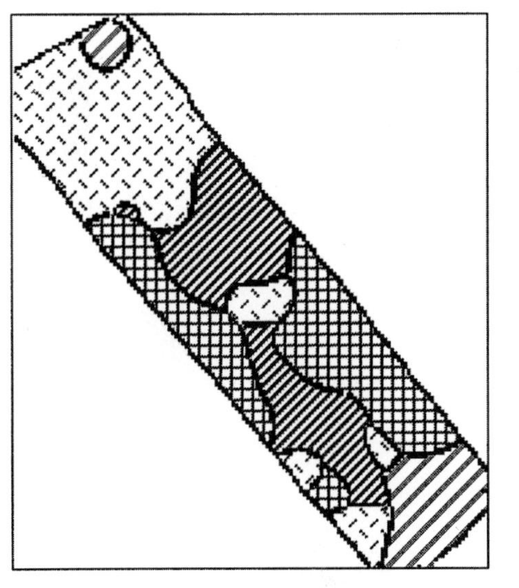

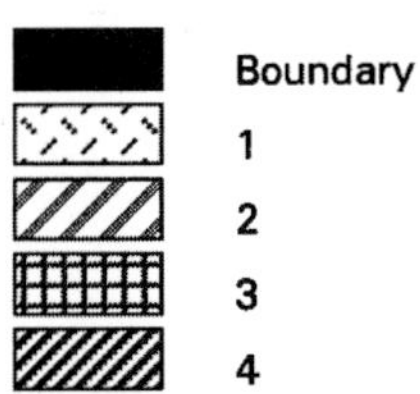

Dominant class assignments
Class 1: low for rape, average for cereals
Class 2: average for rape, low for cereals
Class 3: good for all crops
Class 4: good for rape, average for cereals

FIG. 6. Dominant class assignments calculated from the optimal boundaries between the fuzzy classes. Class 1, low for rape, average for cereals; class 2, average for rape, low for cereals; class 3, good for all crops; class 4, good for rape, average for cereals.

monitor, and possibly to adjust management for, crop growth conditions during a season. The costs to the farmer are then limited to the costs of the sensors and the data analysis package.

References

Bedzek CJ 1981 Pattern recognition with fuzzy objective function algorithms. Plenum Press, New York

Burrough PA, McDonnell RA 1997 Principles of geographical information systems. Oxford University Press, Oxford, in press

Burrough PA 1993 Soil variability: a late 20th century view. Soil Fertil 56:529–562

Burrough PA, van Gaans P, Hootsmans R 1997 Continuous classification in soil survey: spatial correlation, confusion and boundaries. Geoderma 77:115–135

Conway J 1995 Soil mapping, Driffield Bank, Harnhill Farm. In: Davies WPD, Limb M (eds) An introduction to integrated cropping systems. Royal Agricultural College, Cirencester, Gloucester, p 17–21

Davies WPD, Limb M (eds) 1995 An introduction to integrated cropping systems. Royal Agricultural College, Cirencester, Gloucester

de Gruijter JJ, Walfoort DJJ, van Gaans PFM 1997 Continuous soil maps—a fuzzy set approach to bridge the gap between aggregation levels. Geoderma 77:169–195

Dumanski J, Onofrei C 1989 Techniques of crop yield assessment for agricultural land evaluation. Soil Use Manag 5:9–16

Fischer RA 1985 The role of crop simulation models in wheat agronomy. In: Day W, Atkin RK (eds) Wheat growth and modelling. Plenum, New York (NATO ASI Series A86) p 237–255

Kennedy M 1996 The Global Positioning System and GIS. Ann Arbor Press, Chelsea, MI

McBratney AB, de Gruijter JJ 1992 A continuum approach to soil classification by modified fuzzy *k*-means with extra grades. J Soil Sci 43:159–175

Pebesma E 1996 The Gstat package. Department of Physical Geography, University of Utrecht, The Netherlands

Swindell J 1995 Mapping crop yield by use of GPS and GIS. Proceedings 1st Joint European Conference and Exhibition on Geographical Information, The Hague, March 26–31 1995. AKM, Basel, p 409–414

Varcoe VJ 1990 A note on the computer simulation of crop growth in agricultural land evaluation. Soil Use Manag 6:157–159

Vriend SP, van Gaans PFM 1994 FUZNLM, a program for fuzzy k-means clustering and non-linear mapping (modified from a program by Bezdek JC et al). Department of Physical Geography, University of Utrecht, The Netherlands

Wackernagel H, Webster R, Oliver MA 1988 A geostatistical method for segmenting multivariate sequences of soil data. In: Bock HH (ed) Classification and related methods of data analysis. Elsevier Science, Amsterdam, p 641–650

Wesseling C, van Deursen W, Karssenberg D 1996 The PC raster manual. Department of Physical Geography, University of Utrecht, The Netherlands

DISCUSSION

McBratney: Why did you use only 80 of the data points to make your yield maps?

Burrough: Because I have an old FORTRAN program that limits the number of data that can be used.

McBratney: And do you think if you had used 1000 you would still have got the boundaries everywhere?

Burrough: It depends on how you create the surface. If you use an interpolator that produces strong smoothing, like splines or block kriging, the I expect that the contiguous boundaries will be enhanced. If you used conditional simulation to produce a single realization then there would be more short-range variation and then the boundaries would appear everywhere. In this case the 80 sample points showed that the original interpolations of the yield data have a strong spatial correlation, so I am fairly happy that using more points for the fuzzy *k*-means clustering would not have substantially altered the results.

Rabbinge: Does that result in types of boundaries of the systems in space?

Burrough: No it doesn't. What happens is that the really clear boundaries come out. Don't forget that when the soil surveyors are in the field they are also smoothing the soil pattern.

McBratney: I guess what surprises me is that the boundaries actually have a topology: they are actually connected. I can't see how that arises.

Burrough: It surprised me at first, but there are good reasons for this. In a previous paper (Burrough et al 1997) and in Burrough & McDonnell (1997) I showed that when the original data have a strong spatial correlation, then the boundaries come out clearly. This is certainly the case for the Turen soils cited in these works, because there the soils are young and dominated by a topologically controlled alluvial process. I showed you the rainforest composition data (also given in Burrough & McDonnell 1997) because there the original data do not have a strong spatial correlation structure, which is precisely what you find in the great species diversity in a tropical rainforest.

McBratney: When you actually do that thing, how much of the connectivity comes from the final process, which must be some kind of skeletonization?

Burrough: What is important is that the technique is totally independent of the data source. This means that in the course of the year, if you are going over that field with a sensor and collecting more data, you can update your maps very easily: you're not stuck with a particular map. You can also incorporate expert information.

McBratney: I guess the implication here for precision agriculture is as follows. Some people might in the first instance want to create the zones for management and manage these zones uniformly. But then I would say, because I want to go further, I don't want the boundaries, I want the continuous information. But as a first step it is certainly something useful.

Burrough: You have either or both.

Stafford: This is similar to the approach that we developed at Silsoe Research Institute. We have applied our analysis technique to about a dozen fields where yield maps have been available for several years. Generally, the clusters haven't been in contiguous patterns.

Burrough: In this case there is in fact quite a strong relationship between the position in the landscape and yield.

Stafford: The only field where we've seen a clear pattern has shallow soil over chalk.

Burrough: These maps are data driven and they're also use-driven. They're not something that have had boundaries imposed on them. It is a flexible kind of mapping—whether it produces what you want is another matter, but the possibilities are there.

References

Burrough PA, McDonnell RA 1997 Principles of geographical information systems. Oxford University Press, Oxford, in press

Burrough PA, van Gaans P, Hootsmans R 1997 Continuous classification in soil survey: spatial correlation, confusion and boundaries. Geoderma 77:115–135

Uncertainty in hydrogeological modelling

J. Jaime Gómez-Hernández

Department of Hydraulics and Environmental Engineering, Universidad Politécnica de Valencia, 46071 Valencia, Spain

Abstract. Hydrogeological models are built to predict groundwater flow and the fate of contaminants in the subsurface. After the crucial step of building a conceptual model that includes the processes that should be accounted for, parameter values must be assigned to the components of the model. Measured values of these parameters are available only at a few locations, as is the case for transmissivity, hydraulic conductivity or porosity. Therefore, before making predictions about the movement of contaminants in the aquifer, it is necessary to predict the parameter values at unsampled locations. Given the spatial heterogeneity of the parameters involved, this prediction is always uncertain. Model parameter uncertainty propagates to flow-response variables and further to transport predictions. Parameter uncertainty can be modelled using stochastic methods. Stochastic simulation is used for the generation of alternative spatial realizations of the parameter values, which are then used as input to groundwater flow and mass transport models to obtain frequency distributions of the response variables, e.g. flow velocities, arrival times or concentration levels. These frequency distributions help in making risk-qualified decisions. In order to make these frequency distributions as precise and accurate as possible, it is necessary to incorporate all relevant information in the parameter uncertainty model, i.e. it is necessary to condition the parameter realizations to all direct and indirect information. With this aim, new techniques have recently been developed in hydrogeological modelling. One such technique, for the generation of conductivity realizations conditioned to conductivity, piezometric head and geophysical data, is the self-calibrated method.

1997 Precision agriculture: spatial and temporal variability of environmental quality. Wiley, Chichester (Ciba Foundation Symposium 210) p 221–230

Uncertainty is a key component of models in the earth sciences. There are two main types of uncertainty, which play different roles in the modelling process: conceptual uncertainty and parameter uncertainty. Conceptual uncertainty is linked to the conceptual model and therefore is the most consequential in the modelling process. It is due to the imperfect knowledge of the processes being modelled and is very difficult to assess. Once the conceptual model is built, the remaining uncertainty is linked to the mathematical and numerical models and is mostly due to uncertainty in the parameter values and the driving forces of the system.

In groundwater modelling, a typical example of conceptual uncertainty is whether adsorption should be included in the model and, if so, which type of adsorption should be modelled. A typical example of parameter uncertainty is the uncertainty due to the imperfect knowledge of heterogeneous parameters such as hydraulic conductivity, transmissivity, recharge, or uncertainty in the boundary conditions.

Parameter uncertainty

It is parameter uncertainty that this paper is concerned with. More precisely, we will focus on parameters such as transmissivity displaying spatial variability that is not totally erratic but for which there is no deterministic law to explain it. Our final goal is to analyse (and eventually reduce) the uncertainty on predictions of simulation models such us groundwater flow and mass transport models, of which transmissivity is an input parameter.

It is impossible completely to remove the uncertainty on parameter values at unsampled locations. Model builders and users must admit the inherent uncertainty of their models and consequently accept a risk of failure. Once uncertainty is accepted as part of the modelling process, we should determine its sources and adopt a model that will allow us to quantify it and devise strategies to reduce it.

Parameter uncertainty is modelled with a random function model. The aquifer being modelled is unique and unknown, except at sample locations. It is the result of a number of natural processes that cannot be modelled deterministically. This aquifer is then replaced by a set of potential aquifers that could be the result of these natural processes. The uncertainty model consists of an ensemble of aquifers, called realizations, all of them sharing a number of statistical properties that characterize the degree of variability and spatial continuity of the uncertain parameters.

It is important to stress that although the model of uncertainty is a random function model, randomness is not a property of reality, but of the model used to describe it. There is nothing random in reality. The type of predictions that result from the use of probabilistic models are not an image of reality; rather they are tools that aim at reality and help us to learn something about it.

In practice, random function modelling of parameter uncertainty consists of the following steps:

(1) Data collection.
(2) Extraction/decision on the characteristics to be shared by the ensemble of realizations (mean, variance, correlation length, etc.).
(3) Generation of an ensemble of realizations sharing these characteristics.
(4) Analysis of each one of the realizations, for instance, to make transport predictions.
(5) Summarize the results and reach risk-based conclusions.

Conditioning and the self-calibrated method

In order to reduce prediction uncertainty, it is important to include as much information as possible in the random function model. Conditioning is the most powerful technique for uncertainty reduction. A random function model is conditioned if each one of the realizations of the ensemble reproduces the measured data at sample locations. For instance, in two-dimensional groundwater models, transmissivity measurements (direct data) should be preserved at sample locations, and piezometric head measurements (inverse data) should be reproduced by the groundwater flow model applied to the transmissivity realizations. Other relevant, indirect information, such as geophysical and geological data, should also be included in the model.

The self-calibrated method (Gómez-Hernández et al 1997a,b) is a Monte-Carlo technique that allows the generation of transmissivity realizations conditional to transmissivity data, piezometric head data and geophysical data. The inclusion of piezometric head data in the generation of transmissivity realizations is a problem that has only recently been solved adequately. The self-calibrated method is used next to demonstrate how increasing the amount of conditioning information can help in improving groundwater model predictions. A multi-Gaussian random

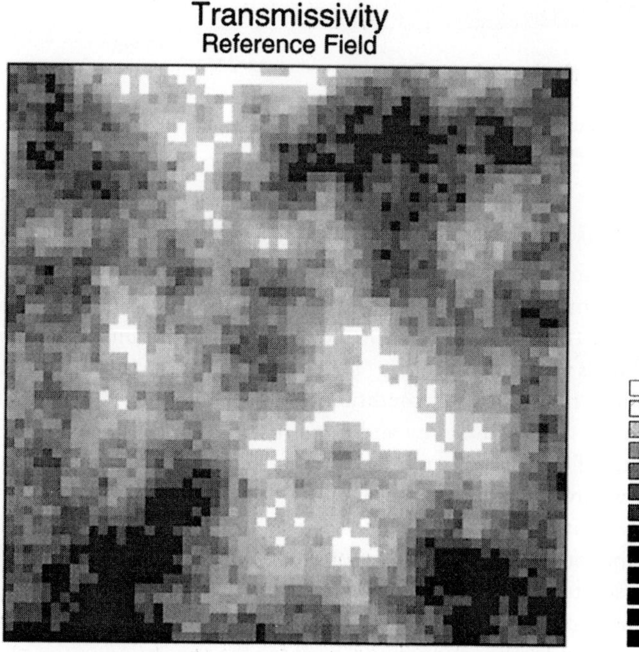

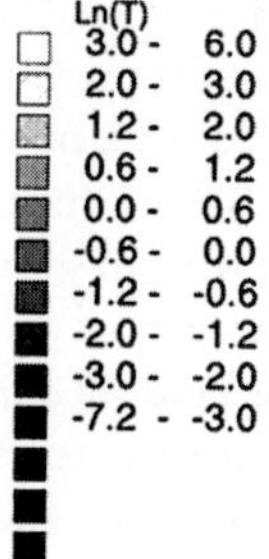

FIG. 1. Realization of log transmissivity used as synthetic aquifer.

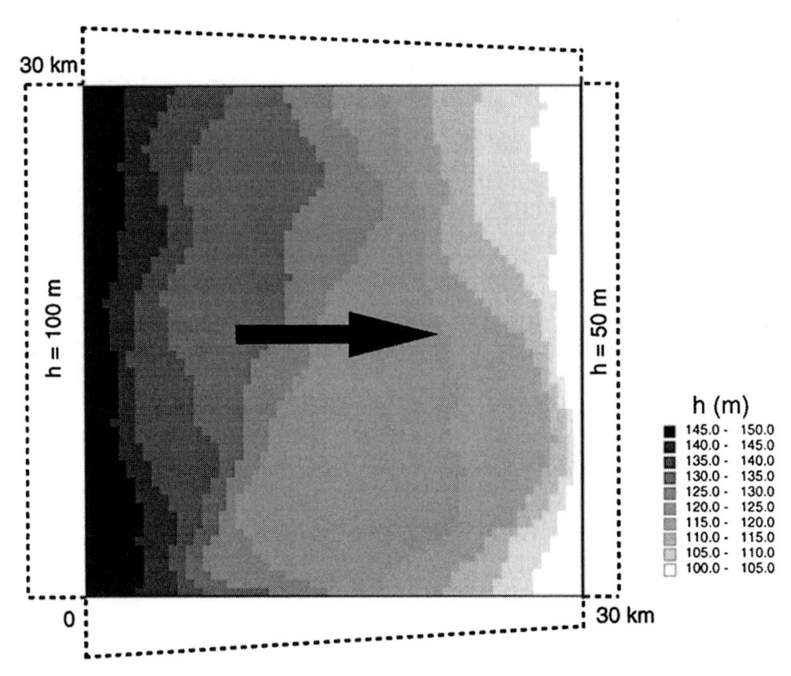

FIG. 2. Boundary conditions used for the solution of the groundwater flow equation, and solution in the synthetic aquifer.

function model is adopted to describe the spatial variability of log transmissivity (logT), within a square domain, with zero mean logT, unit variance and isotropic correlation length of one-third of the size of the simulation domain. One realization from this random function is generated using sequential Gaussian simulation (Deutsch & Journel 1992) and is taken as reality (Fig. 1). Groundwater flow is solved within the domain for prescribed head boundary conditions as shown in Fig. 2. This realization is sampled for a number of transmissivity and piezometric head data at seven locations (Fig. 3). The impact of conditioning to each type of data is described next.

We wish to show how increasing the amount of conditioning data helps to improve mass transport predictions. For this purpose we need to establish the global statistics of the random function model, the quality of which will obviously depend on the amount of data used to obtain them. We will assume the same global statistics—therefore the same random function model—for all the sets of realizations, and equal to those of the random function model used for the generation of the synthetic reality. In this way we are using the best estimates possible for the global statistics. The

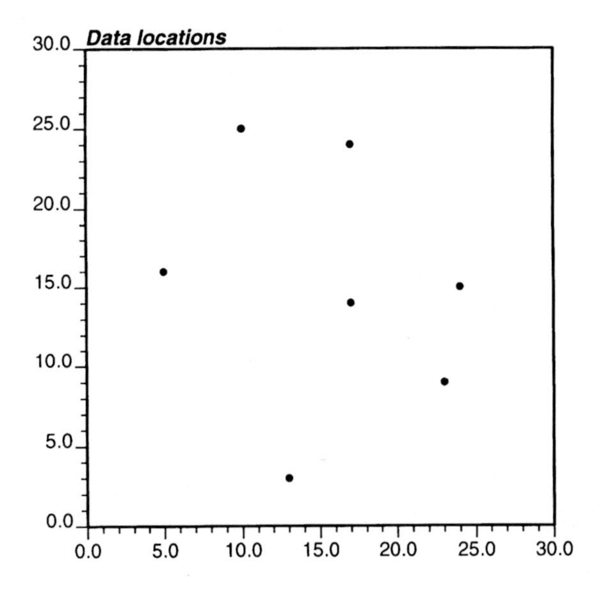

FIG. 3. Sampling locations for head and transmissivity data.

results will show the improvement due to just considering different sets of conditioning data. If the global statistics of the random function model were to be inferred from the sample data, the improvement in the prediction quality will be larger since the quality of the global statistics estimation will also increase with the amount of data.

Four cases are analysed: (1) neither transmissivity nor head data are available, unconditional realizations are generated respecting only the global statistics of logT, i.e. zero mean logT, unit variance and isotropic correlation length equal to one-third of the simulation domain; (2) only transmissivity data are available; (3) only head data are available; and (4) both transmissivity and head data are available. In each of the four cases, 100 realizations are generated, groundwater flow is solved, and a single particle is tracked from the centre of the upstream boundary until it leaves the flow domain. Figure 4 shows the paths of the particle as it travels across the 100 realizations for each of the four cases. A clear reduction of the uncertainty on the particle trajectory can be noticed. The largest reduction occurs when both head and transmissivity data are used. Conditioning to only head or only transmissivity also reduces the uncertainty with respect to the unconditional case, each type of data in a slightly different way.

The reduction of uncertainty can be quantified by analysing the spread of the particle paths as they cross selected control planes, or by analysing the distribution of travel times (Wen et al 1996, Gómez-Hernández & Wen 1994). For instance, Fig. 5 shows a

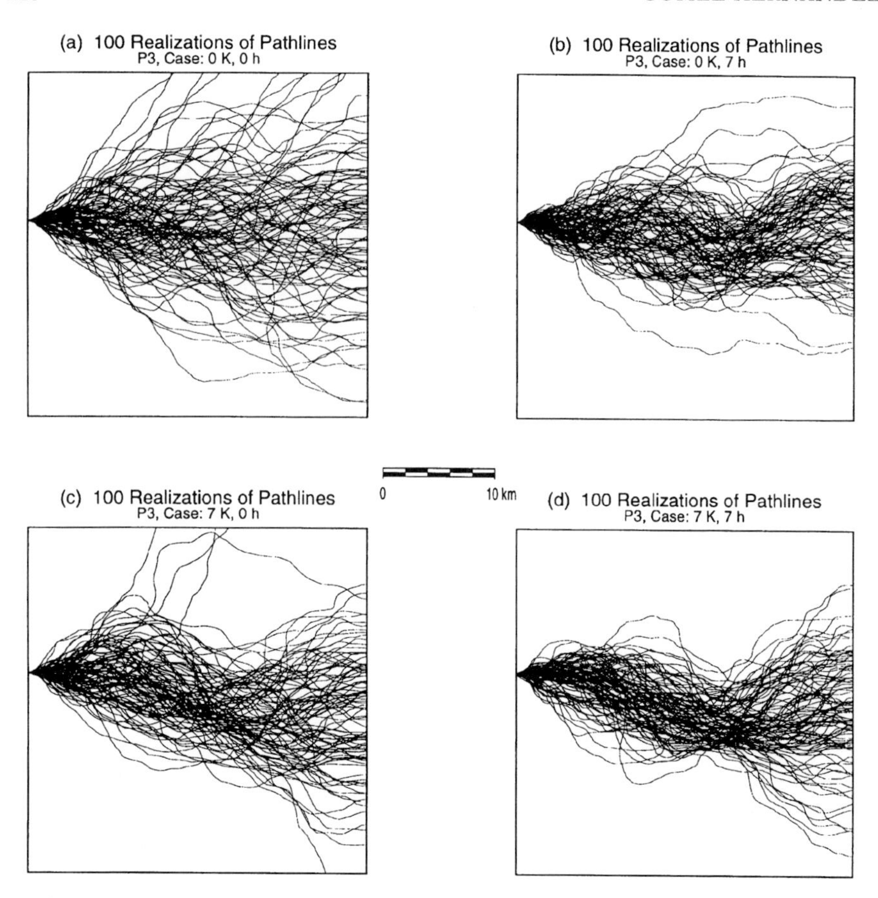

FIG. 4. Trajectories of a single particle in the 100 realizations corresponding to each of the four random function models used.

contour plot of the residual uncertainty after conditioning to a certain combination of head and transmissivity data. Each point in this figure indicates the uncertainty on the arrival position of the particle for a given combination of number of head data and number of transmissivity data, expressed as a ratio with respect to the uncertainty for the unconditional case. For instance, for a combination of 25 transmissivity measurements and five head measurements, in this particular exercise, the uncertainty in the arrival position, as measured by the variance of the arrival location, is reduced to 50% of the uncertainty from the unconditional case. This figure displays the trade-off existing between head data and transmissivity data and could be used, for instance, in a network design project.

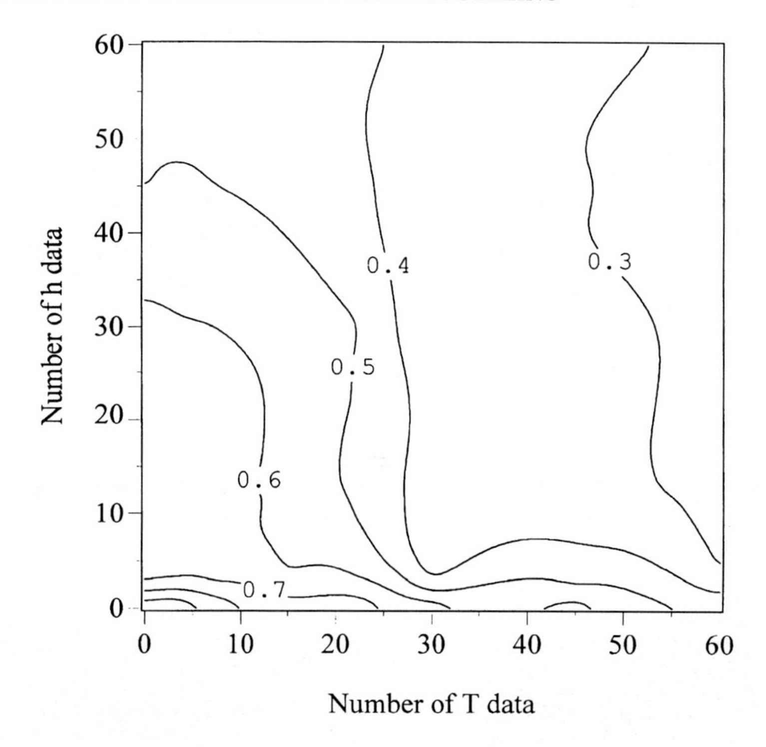

FIG 5. Ratio of the uncertainty in the arrival position as obtained by different combinations of head and transmissivity data with respect to the unconditional case.

Summary

We must recognize that the predictions of our models are uncertain. Then, we can approach the groundwater flow and mass transport modelling with the objective of making risk-qualified predictions, since all predictions will be accompanied by an uncertainty value. Injecting as much information as possible into our uncertainty models will reduce the magnitude of the uncertainty estimates. For this purpose, the self-calibrated model has been shown as a powerful tool for incorporating transmissivity, piezometric head and geophysical data in our transmissivity models.

References

Deutsch CV, Journel AG 1992 GSLIB: geostatistical software library and user's guide. Oxford University Press, New York

Gómez-Hernández JJ, Wen X-H 1994 Probabilistic assessment of travel times in groundwater modeling. J Stochastic Hydrol Hydraulics 8:19–56

Gómez-Hernández JJ, Sahuquillo A, Capilla JE 1997a Stochastic simulation of transmissivity fields conditional to both transmissivity and piezometric data, 1: Theory. J Hydrol, in press
Gómez-Hernández JJ, Capilla JE, Sahuquillo A 1997b Inverse conditional simulation. In: Baafi EY, Schofield NA (eds) Geostatistics Wollongong '96, volume 1. Kluwer Academic, Amsterdam, p 282–291
Wen X-H, Gómez-Hernández JJ, Capilla JE, Sahuquillo A 1996 Significance of conditioning on piezometric head data for prediction of mass transport in groundwater modeling. Math Geology 28:951–968

DISCUSSION

Rabbinge: We have seen how we can decrease the uncertainty which we have to work with. Procedures such as those described in this paper are extremely valuable because otherwise we end up simply saying we do not know, whereas with this technique we can know what it is we do not know. This is important for future decision making, and for the process through which we develop better precision agriculture.

Braud: You have only one parameter distributed in space in your example, whereas in precision agriculture we are dealing with models with several parameters which are often numerous and have different probability distributions. Do you think that your technique can also be used in this case, and is it easy to do so without too much computing time?

Gómez-Hernández: We actually deal with more than one parameter in hydrogeology, too. For example, we can account for the joint spatial variability of hydraulic conductivity, porosity, aquifer thickness and recharge.

Braud: And for the generation of the realization, do you have techniques which you can take into account a possible correlation between parameters?

Gómez-Hernández: There are algorithms readily available for the generation of joint realizations of multiple parameters, accounting for the auto-correlation of each parameter and the cross-correlation among them.

Stein: I was particularly impressed with your use of the soft data. We have some experience with soft environmental data. You can combine data on environmental variables quite easily with soft data which describe whether you exceed environmental thresholds, but which do not express the actual content. How do you combine the soft data with your simulations?

Gómez-Hernández: In this case we used an indicator approach. In this approach each datum must be transformed into a series of indicator variables. An indicator variable for a given threshold takes the value of 1 if the datum is below the threshold, 0 if it is above. The first step is to discretize the range of values of the hard variable into a number of thresholds (usually in the order of 10). The hard data are simple to transform into the series of indicators just by comparing the value to the different thresholds selected. The soft data require some kind of calibration information from which one can decide whether the hard datum will be above or below a given threshold knowing the value of the soft datum. Soft data generally become a set of incomplete series of indicators, since there is an area in the range of the hard data in which knowing the soft datum value is not enough to discriminate the corresponding hard variable value. Later, during the estimation or stochastic simulation steps, the gaps in the series of indicators corresponding to soft data are filled.

Stein: The choice for the soft data is rather critical, if you want to get a good reduction in the uncertainty values.

Gómez-Hernández: The soft data must carry enough information in the hard variable to allow you to discriminate if the hard variable is above or below some of the thresholds used to discretize the range of the hard variable. The fewer incomplete indicator variables in the series left, the more precise the soft datum is, therefore the better. One example in soil science is soil texture. Soil texture can be used to constrain the range of variability in hydraulic conductivity, for instance.

Rabbinge: So what you are saying is that if the soft information you use to reduce the uncertainty are to be made more accessible for kriging, then you must first know some more basic information which you bring in. For instance, soil science, texture and hydrology are probably characteristics related to particular phenomena.

Gómez-Hernández: There are many ways of using soft data in geostatistics. Let's consider a scenario in which the hard variable is hydraulic conductivity and the soft variable is soil texture. One can make a regression plot of these two variables from a calibration data set and from the regression cloud read the range of conductivity values corresponding to each texture. From this information, the soil textures can be transformed in constraint intervals that can then transformed into incomplete series of indicators.

McBratney: In this case, I think the technique and the software work well because the soft data contains the topology of the system, because of the detail. The hard data consists of point information and doesn't really give you that topology. Topology is rather important in this physical example. Of course in the model that you were using, the physics was not uncertain, whereas yesterday Vic Barnett was telling us that many crop simulation models were rather uncertain (Barnett et al 1997). We really have to deal with both of those issues and that we don't know which one of these is larger at the moment, and the situation is even more uncertain.

Gómez-Hernández: I did mention at the beginning of my talk that there are two sorts of uncertainty that we are dealing with: conceptual uncertainty, which is always large and difficult to quantify, and parameter uncertainty. In my talk I focused on parameter uncertainty, and I agree with you that in hydrogeology, conceptual uncertainty in flow modelling is generally low (the process of subsurface flow is well understood), and that in the type of models used in precision agriculture, conceptual uncertainty may play a much larger role than parameter uncertainty. I must add that I decided to use a simpler example in my presentation than the one included in the paper. For those interested, this example is taken from Gómez-Hernández & Wen (1994).

Rabbinge: That's why in many of the simulation studies, the concepts are not clear. But if you use the model to refine the concepts, or to criticize them, then it's an iterative procedure with a high heuristic value.

References

Gómez-Hernández JJ, Wen X-H 1994 Probabilistic assessment of travel times in groundwater modeling. J Stochastic Hydrol Hydraulics 8:19–56

Barnett V, Landau S, Colls JJ, Craigon J, Mitchell RAC, Payne RW 1997 Predicting wheat yields: the search for valid and precise models. In: Precision agriculture: spatial and temporal variability of environmental quality. Wiley, Chichester (Ciba Found Symp 210) p 79–99

General reflections

Johan Bouma

Department of Soil Science and Geology, PO Box 37, Wageningen Agricultural University, 6700 AA Wageningen, The Netherlands

Defining sub areas within fields

My goal here is to try to outline a framework within which we can fit the various activities we have talked about during this symposium. Our reference point is the classical agricultural practice of uniform treatment of fields. In precision agriculture we take fields and define sub areas within them, which we then manage differentially according to their properties. During this meeting we have heard that there are four methods for defining these sub areas.

(1) The classical soil survey, in which the pedogenic information has to be translated into a functional form. Here we include landscape analysis and digital terrain modelling.
(2) On-the-go harvest patterns for consecutive years (yield maps).
(3) Remote-sensing patterns, which can also be analysed in time throughout the growing season when multitemporal images are available.
(4) Simulation modelling of important land qualities, relating to factors such as water movement, crop growth and agrochemical fluxes. These use interpolation techniques to obtain areas.

In all cases we end up with various patterns in time and space which have to be analysed and compared. Here is to be found a major role for geostatistics. I am not suggesting that these are entirely different approaches: there are ways of combining them. In the future, I envisage the development of a protocol in which these methods are neatly combined.

What are the overall goals?

In precision agriculture it is vital that we define our goals. It is worth emphasizing that these are not really our goals at all, but rather those of the farmers, our stakeholders. These include profitability, product quality, risk reduction and environmental protection. In the latter case we are beginning to see laws defining specific limits and threshold values for the operator: this will increasingly occur. We have heard from

Pierre Robert that 80% of sugar beet farmers in Minnesota are using site-specific management because by playing around with the nitrogen regime they can significantly improve product quality, which leads to increased profitability. Daan Goense has told us about the work in France, part of our EU project in precision agriculture that he is coordinating (John Stafford is also participating in this) where they were harvesting barley on a field. Half the field produced barley of malting quality, the other half didn't. This has to do with factors such as water regimes and nitrogen manipulation. The whole crop ended up being sold at a low price. You can do two things here: (1) harvest the different quality crops separately; or (2) try to improve the half of the field that didn't produce malt-quality barley. Another example has to do with the starch quality of potatoes. It is important that these examples are being highlighted.

How should we handle legislation for environmental thresholds of agrochemicals? I liked the example Peter Groffman gave, where in the USA car manufacturers were told to make cars more economical: the government set the target level ('performance' rather than 'design' indicators) and left it up to the manufacturers to decide how they were going to meet it. In agriculture, we should give responsibility to the people who deserve to have it and who can handle it — our modern farmers. We, as scientists, have a major responsibility to define what it means to have environmental threshold values in a highly variable medium in space and time, as opposed to the industrial criteria. This whole idea of moving towards performance indicators is a very good one and may feed our interaction with our users. There is a major role here for statistical procedures (e.g. Vehagen & Bouma 1997).

The farmer's toolkit

If these are the goals for the farmer, what is his toolkit — that is, the key elements of the production system — which he can use to achieve these goals? He can:

(1) Manipulate his nutrient management (N, K, P) by using manure and fertilizers.
(2) Control pests and weeds by biocide application and by using integrated pest management techniques.
(3) Manipulate the structure of the soil by using various tillage practices, including zero tillage which has favourable implications for erosion (if you can take care of the problem of weeds).
(4) Use different crop varieties, for instance by planting an early variety on a dry spot and a late variety on a wet spot.

All these elements constitute the toolkit which becomes part of the decision support system: 'what' to do 'when' and 'where'. Traditional agriculture already involves controlling management in time (e.g. split applications of fertilizer); precision agriculture adds to this differential management in space.

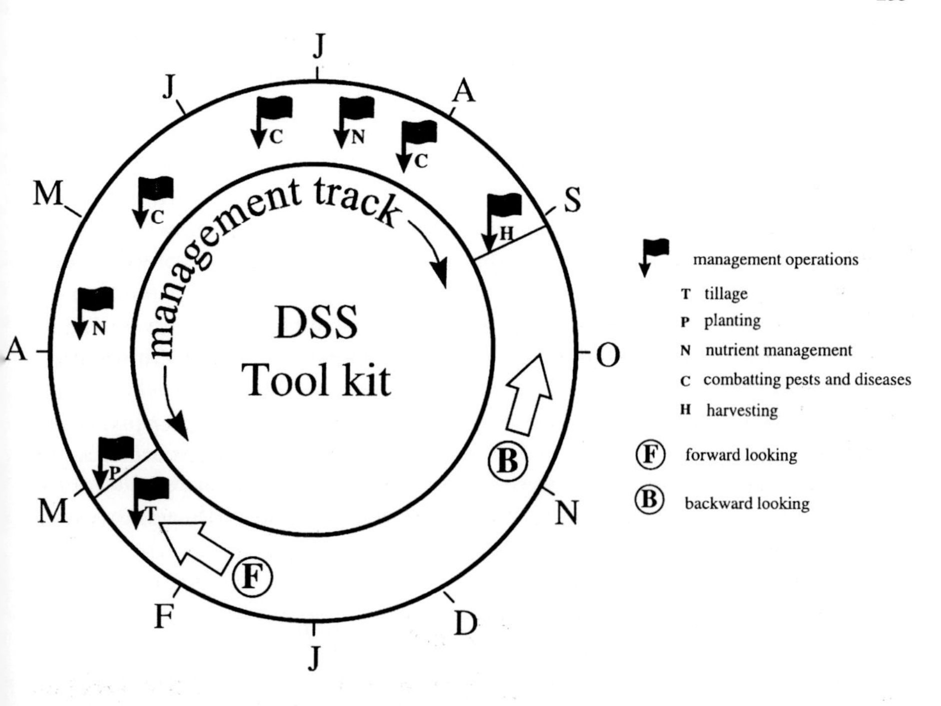

FIG. 1. Management toolkit for the 'management' track for potatoes. The management track is part of preceding and following land use.

Figure 1 shows what we can call the 'management track' for precision farming. This is actually a management track for a potato field in the Netherlands. The flags represent the items from the toolkit: elements of the production system that can be applied at different times and locations. The management track is part of a continuous cycle including other periods of the year. The question is, how and when is the farmer going to intervene, and how can we help him in his decision making? How much does he benefit if we do this site-specifically? We are talking about the fine-tuning of management.

How can differences within fields be observed?

With regard to how a farmer can alter the production system, there are two possibilities. The first is the reactive approach of on-the-go monitoring and application of agrochemicals using sensors, which is a technique that will probably become more important in the future. The more traditional method involves sampling sub areas of the field and basing management on this. We have heard a lot about sampling this week, including targeted sampling, adaptive sampling and ranked set sampling. This sampling can either be within the soil or by remote sensing.

When will precision management take place?

The second issue concerns the timing of these interventions: when is each element of the tool kit applied? Currently this happens on the basis of the expert knowledge of the farmer. One thing that has come out of this meeting is the possible role of space–time and regression analyses, looking at production data in the past as a function of production factors. This kind of information can be used to target interventions in time. The other way of doing this is by means of exploratory simulation modelling, which we have been able to focus on the water regimes, from which many other factors are derived. We can simulate these moisture regimes pretty well, also in a predictive mode (e.g. Booltink & Verhagen 1997). This is not only for one-dimensional soils; it is also for landscapes. Uncertainty analysis needs to be part of this and there is also a link here with the metadatabase. Geographical information systems (GIS) are extremely important. Again, a major role for statistics is foreseen.

Combining the 'what?', 'where?' and 'when?' questions

What I would like to stress in this scheme are the basic activities and their sequence. We start in the spring, and we don't know what is going to happen with the weather: we are therefore taking the forward-looking approach. Weather is extremely important. We have a soil database of soil properties. The idea is that we can calculate certain key characteristics of soils and crops which suggest to us when to go in with one of our tools. This way we can be proactive and don't have to wait until remote sensing tells us that the crop is in trouble, which would mean that some damage has already been done. At the end of the growing season we should also be backward-looking. Many farmers are interested in playing 'management games' in winter when they have time. The challenge for us is to demonstrate that the use of the precision agriculture toolkit gives better results than traditional management procedures. I'm not sure what the decision support system will finally look like, but I am convinced that in this scheme, the farmer comes first. Anything we do will ultimately be evaluated according to the four goals of the farmer I mentioned earlier.

The role of research

A statement we hear increasingly from policy makers is that after 100 years of agronomic research we now know most of the answers. Is new research for precision agriculture really needed or can we combine existing knowledge from different disciplines? Research is needed on sampling, on error propagation during modelling and on integration of models and expert systems in a GIS environment. Much work is also needed to develop decision support systems that allow a systematic consideration of many contrasting requirements and demands that confront the farmer. Precision agriculture requires truly interdisciplinary research in which every discipline has to reconsider its role (Bouma 1997). Without exaggeration we may conclude that a new research paradigm is emerging.

References

Booltink HWG, Verhagen J 1997 Using decision support systems to optimize barley management on spatial variable soil. In: Kropff MJ, Teng PS, Aggerwal PK et al (eds) Applications of systems approaches at the field level. Kluwer Academic Publishers, Dordrecht, The Netherlands, p 219–235

Bouma J 1997 Role of quantitative approaches in soil science when interacting with stakeholders. Geoderma, in press

Verhagen J, Bouma J 1997 Defining threshold values at field level: a matter of space and time. Geoderma (special issue), in press

DISCUSSION

McBratney: I have always maintained that it is important for us to know the historical scientific literature. With this in mind, I would like to point out that this is not the first time we have had a revolution like precision agriculture: a similar one occurred in 1731. The question I am posing now is: could we have had this meeting 265 years ago and come up with the same conclusions? Figure 1(*McBratney*) shows Jethro Tull and the

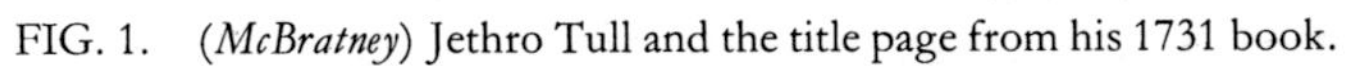

THE NEW
Horſe-Houghing Husbandry:
OR, AN
ESSAY
ON THE
PRINCIPLES
OF
Tillage *and* Vegetation.

Wherein is Shewn,
A METHOD of INTRODUCING
A Sort of
Vineyard-Culture into the *Corn-Fields*,
In Order to
INCREASE their PRODUCT, and DIMINISH
the COMMON EXPENCE,
By the USE of
INSTRUMENTS lately Invented.

LONDON:
Printed for the AUTHOR, in the Year M.DCC.XXXI.

FIG. 1. (*McBratney*) Jethro Tull and the title page from his 1731 book.

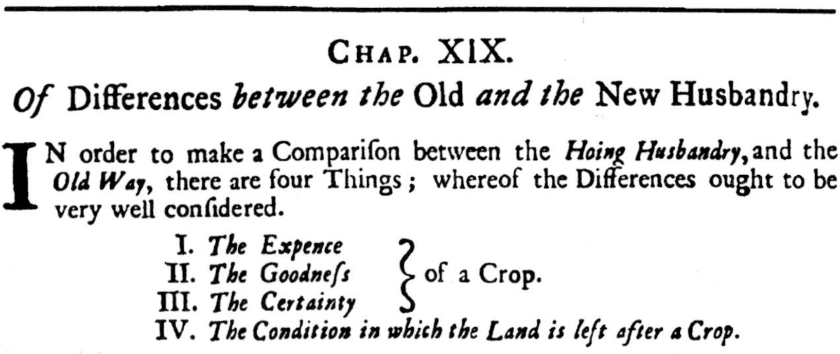

CHAP. XIX.

Of Differences *between the* Old *and the* New Husbandry.

IN order to make a Comparifon between the *Hoing Husbandry*, and the *Old Way*, there are four Things; whereof the Differences ought to be very well confidered.

I. *The Expence*
II. *The Goodnefs* } of a Crop.
III. *The Certainty*
IV. *The Condition in which the Land is left after a Crop.*

FIG. 2. (*McBratney*) Chapter heading from Jethro Tull's 1731 book *Essay on the principles of tillage and vegetation*.

title page of his book about growing things in row crops and dealing with weed management. His aimed to 'Increase the product, and diminish the common expence (*sic*), by the use of instruments lately invented'. I see a clear parallel with precision agriculture. Moreover, what is amazing is that the goals Johan Bouma talked about—profitability, product quality, risk reduction and protection of the environment—were published way back in 1731! Figure 2 (*McBratney*) testifies to Tull's prescience. Here Tull states that he is going to compare the old way with the new way, and to do that he has to show the following considerations: "The expense [the profitability], the goodness [the quality], the certainty [the risk] and the condition in which the land is left after a crop [the environment]". The same criteria and, amazingly, in the same order as Johan Bouma has just expounded. People were just as clever in the past: although they might not have been able to achieve their goals, they knew what they wanted to do.

Groffman: The management track Johan Bouma describes illustrates the strengths of precision agriculture, but it also shows the limitations. The track only addresses half the year. We know that the water, nutrient and soil losses occur in the other six months of the year. Unless we're prepared to address the rest of the year when the losses occur, and the larger context of the system in terms of its sustainability, then we're still going to have problems with sustainability and environmental quality. The management track is an excellent place to put precision management in the context of a larger tool kit, but let's be clear about it: the tool kit has to include things other than those which contribute to yield.

Bouma: I fully agree. I considered the environment as well by considering leaching. I used this track only as an illustration of growing one potato crop. However, the entire year is important as leaching occurs mainly in the wet season when the soil nitrogen content is too high (e.g. Booltink & Vehagen 1997).

Burrough: Are farmers generally aware of processes such as leaching that are going on in winter in the same way as they are when the crop is growing? If they are not, would the use of visualization tools help to make them more aware?

Groffman: Farmers are very aware of trafficability and erosion. But farmers still till throughout the year. A high proportion of agricultural land suffers from compaction and poor soil structures because even though farmers know they shouldn't till and harvest when the soil is wet, they do it because they have to. That illustrates another problem: farmers know a lot about how their system works but they have a conflict between short-term needs and long-term degradation.

There's been a lot of work done in the USA concerning the extent to which farmers recognize and respond to environmental problems. The results show that they seem to have a pretty good awareness of the way systems work, and they seem to be willing to do things to make them work better as long as it doesn't put them out of business.

Burrough: We had some information in Marc Voltz's paper about the nutrients that were picked up in the drains and didn't leave the catchments (Voltz 1997, this volume). Were your farmers aware that this was happening?

Voltz: To be frank, we didn't tell them, because if we had told them that pesticides were leaching from the fields they would have stopped cooperating with us. We have difficulty getting farmers to participate in our research programmes, especially concerning farming pollution. Sometimes they'd rather not know, because they are afraid that society will then saddle them with crippling environmental constraints. In northern Europe this is less of a problem because the environmental concern is stronger, but in southern Europe it's rather difficult.

Bouma: That is an important comment. I would point out that manipulating soil structure can be an effective way of changing flow patterns of water in soil. 'Tillage' is part of the toolkit. Often pollution of groundwater by pesticides is caused by bypass flow, water rapidly following large vertical pores in the soil. You can manipulate your soil structure by tillage at certain points in time so that the flow patterns go through the soil matrix and agrochemicals are filtered out.

Voltz: I agree with that, but we must start by explaining that there is a problem, and that it is in the interests of the farmer to help us. But it's not easy.

I have a comment about the variability of farming strategies in different regions of the world, and the different categories of farmers. From time to time during this symposium I have had the impression that we are only speaking about one kind of farming system, involving large fields. There are also other farming systems with small fields. These farmers are not what we could call model farmers, and their strategy is often very different. We have to adapt our precision farming techniques to all kinds of farming strategies.

Groffman: A concept that seems to be useful for communicating to farmers is that of mass balance. Farmers have a pretty good sense of the nutrients that they put in and a good sense of what comes out at harvest. You can calculate the balance when you know that what goes in has to come out. It either has to come out of the crop, or in the water or the air. This concept originated in The Netherlands, in the 1977 book 'Cycling of mineral nutrients in agroecosystems' (Frissell et al 1977).

Burrough: The modelling behind the visualization is mass balance modelling.

Bregt: A general comment on the framework Johan Bouma proposed. He mentioned that in the past there was extensive research on traffickability which was not actually used by the farmers. We were also involved in developing decision support systems, not for precision farming but for physical planning. Our experience was that we did a nice job, but the work was not used. This was because we didn't listen to the questions or real problems of the users. Now we are starting again and I hope we will be more successful. First we will do an intensive analysis of the real needs of the people who are to use these systems. What we have been missing in our discussions is that we should really be thinking of systems that are to be used by farmers.

Mulla: There is an excellent point in Johan's diagram which I would like to make explicit. This is the need for researchers to work in a multidisciplinary setting. When we do field research it should be large in scale, long in time and it should encompass measurements of many different facets of the system. It is the system that the farmer is managing—the tillage, the weeds, the insects, the fertility, the crop yield and the environment. We, as researchers, often go to the field and measure soil fertility without measuring yield, or yield without measuring weeds, or landscape attributes without measuring water. We have to recognize that the farm is actually an ecosystem by itself: it has many components, and precision farming must deal with the interaction of all of those components as a system, otherwise it will fail.

Kropff: I think these kind of systems approaches are important. But in terms of choosing between working in the laboratory and working with farmers, we have to be careful. In the agricultural sciences there have been cycles where people have begun to work with farmers solving problems and then they have moved into climate chambers to study detailed processes, and then 10 years later they moved back out into the field. In Wageningen we are developing a programme that is focused on the development and design of new integrated and ecological farming systems which achieve multiple objectives. Some scientists are working with farmers and others are doing fundamental research to solve specific problems that can only be addressed on experimental farms or in climate chambers. We have found that it is very stimulating to link these different research activities so that a knowledge chain is developed between applied and fundamental sciences.

Goense: I agree with what David Mulla and Martin Kropff have just said, but I think we must go even further, towards a total farm system. It is not only the factors within the field we have to consider, but also the organization of the farmers work. What might be expected of us as researchers is for us to translate and analyse the complex systems within which farmers work. A further aspect is that we must deal with the people who have to implement the science, i.e. the agricultural software houses. We as scientists must make clear how the total system is organized.

Rabbinge: I would like us to discuss a subject that has been raised extensively during our meeting, namely how to improve tools such as sampling, geostatistics and modelling.

Stein: As far as I am aware in this meeting we have, for the first time, brought together a group of high-level statisticians, agriculturalists and scientists who think that there is some unifying concept in precision agriculture. Over the last few years, large advances have been made in sampling techniques, for instance adaptive sampling. Dealing with variation in space and time is a much more recent topic, facing many difficulties. But still much progress can be expected. Space–time analysis is governed much more by data availability and the required specific analysis is much more crucial in this kind of work than before. Further, we need more procedures to deal appropriately with elements of scale as well as methods for comparing spatial patterns, for example, patterns of yield, related to soil and hydrological patterns, as shown by Verhagen and Bouma (1997).

Rasch: Concerning sampling, we need more techniques for sampling in space and time at the same time. It is also necessary to take costs into account: both the cost of sampling and also the cost of a wrong decision due to too-small a sample size. We need some maximum risk we can accept, and from all this we have to decide the size of our samples. How many observations do we need to draw conclusions from?

Barnett: I am glad there has been an emphasis on sampling in this meeting and I echo all that has been said. A number of modern techniques are being honed now and are being widely used in the environmental field. They have the characteristic of moving us away from the direct random sampling to use more information about the structure that we are sampling, so that we can be much more efficient. I do not think I would want to make any prescriptions about sample sizes. I am reminded of a recent exercise that I was involved in which had to do with public utilities and the need to evaluate long-term development costs in the water industry, where every single observation that was taken was likely to cost a minimum of US$500 000. When observations cost that much you do not talk in terms of taking a minimum of 10 observations.

The point behind this remark is that there is quite a strong move in certain areas (which is not I believe reflected in the precision agriculture field) towards Bayesian sampling. In this, you seem to be getting something for nothing, even more than you are in the ranked set sampling which worried Peter Groffman yesterday. It is an even more extreme version of that, which involves making extensive use of so-called expert knowledge. A lot of the investment policy in the water industry for long-term development is based nowadays totally on Bayesian sampling: that is, on choosing very small number of observations very much guided by what the 'professionals' (and I use this word in the widest possible sense) combine to know and present about the situation. In this way sample sizes have become very small and certainly people believe they are getting accurate results. But I return to my first point of thoroughly welcoming the discussion of sampling within this context here. I think it is a vital component, as indeed is modelling (in spite of what you may have thought from what I said earlier in the meeting), because the key to everything in precision agriculture is that we must, as in all walks of life, look and learn. We look to see what is happening and we try to learn from what we see. This means that all of us have to be modellers: we have to try to represent what is going on. I am totally behind the contribution of all aspects of information to modelling. This includes empirical information and the scientific input into these models. After all, there would be no point in just constructing an empirical model and ignoring what you know about the science that underlies it. I do not want to give the impression that my contributions here were nihilistic: they were, if anything, a little sad in encountering one area where an enormous amount of investment has gone into model development that has not achieved what many people think it should have achieved and are trying to make it achieve.

My final point is a precaution that I offer to even my most junior students, which concerns the meaning of the word 'simulation'. Simulation is to do with the pseudo-random replication of what happens in a physical situation. You hope to be able to produce a sample of information from the world situation that you are looking at. The extent to which you are successful in this must depend on the extent to which you understand that situation, but too many people hide behind what they call 'simulation models' which really are based on no information at all: they are handle-turning. They throw up some data and say that is what would happen in this situation. This is fine if it's based on good science. But we must discourage people from hiding behind the term 'simulation modelling'. We must urge against exhortations to use simulation sampling other than in a situation where you really are simulating—i.e. representing truly what goes on in real life.

McBratney: In precision agriculture there is a kind of sampling that we will have to adopt that consists of monitoring networks to record things like soil moisture, because we're going to need that temporal information. We haven't really talked about how one strategically sets up those kinds of networks. These devices are still not very cheap, although they're coming down in price: you're still talking about a few thousand dollars per site to set up a recording device to measure soil moisture. Then there's the type of sampling where we are not so much worried about monitoring, but involve spatial sampling in the field where we're collecting information on nutrients and weeds and so on. Then we can use some of our fancy techniques to direct the sampling, but still we haven't talked about the economics of that. Is the economics of that reasonable? The line of research that we're following is to say that perhaps in the long run it's going to be too expensive and therefore the way to do this is to have a few measurements and lots of less precise measurements but with much more spatial precision. This can be achieved by scanning technologies, proximal sensing as I call it, and we really didn't talk about that very much. These are a few approaches that we have to develop along with the statistics.

Thompson: We want sampling strategies that give good estimates or allow us to make useful maps, but at the same time the procedures have to be simple and economically feasible. From a practical point of view I think that double and multiphase sampling designs may become increasingly useful. With these strategies a large amount of less precise data, collected for example by remote sensing or scanning devices, is combined with more precise data, such as exact measurements made on the ground.

McBratney: One of the problems with remote sensing has been that farmers need the information quickly, and in the past they haven't been able to get it the next day.

Mulla: I'm organizing a symposium for the 1997 Soil Science Society Annual meeting at Anaheim, CA, entitled 'Remote Sensing for Precision Agriculture'. Some of these issues about delivery of information in a timely fashion will be addressed explicitly.

Burrough: There is an intimate relationship between the sample spacing and the scale of the patterns you are trying to pick up, and this is shown very nicely by Salvador Dalí in his painting 'Gala contemplating the Mediterranean Sea which at twenty meters becomes the portrait of Abraham Lincoln'. As you get further away you go from one pattern into another. At one scale you're observing the fine detail of the young lady in the window, but as you get further away you see the head of Abraham Lincoln. How many times when we're sampling something at one scale do we actually try to extrapolate that to another scale? If you're taking a small number of data points close in the field you're looking at that local pattern, you're not looking at a global pattern. If you want to look at the global pattern you have to sample it a different way.

Barnett: A brief remark about remote sensing: just a plea for some research to be done on what to measure from the remote sensing. At the moment there seems to be a rather *ad hoc* use of two-band responses in the form of simple ratio or difference. Since remote sensing started, there has been a desperate need to analyse the most informative ways of handling the signals that you get from different channels.

Webster: I would like to put in a plea for more research on robust sensors that we can put in the soil. They have got to be robust and cheap so that we can have dense information, and they must respond quickly so we can feed the information to farmers for operation the next day. This is absolutely necessary if we are to make use of the ideas that we have for precision operation.

References

Booltink HWG, Verhagen J 1997 Using decision support systems to optimize barley management on spatial variable soil. In: Kropff MJ, Teng PS, Aggerwal PK et al (eds) Applications of systems approaches at the field level. Kluwer Academic Publishers, Dordrecht, The Netherlands, p 219–235

Frissell MT, Hoate S, Newbould P (eds) 1977 Cycling on mineral nutrients in agroecosystems. Elsevier Science Publishers, Amsterdam

Verhagen J, Bouma J 1997 Defining threshold values at field level: a matter of space and time. Geoderma (special issue), in press

Voltz M 1997 Spatial variability of soil moisture regimes at different scales: implications in the context of precision agriculture. In: Precision agriculture: spatial and temporal variability of environmental quality. Wiley, Chichetser (Ciba Found Symp 210), p 18–37

Summary

Rudy Rabbinge

Scientific Council for Government Policy, 2 Plein 1813, P.O. Box 20004, 2500 EA, The Hague, The Netherlands

In summarizing, I would like to make five major points. First, agriculture and agricultural sciences are evolving and precision agriculture is part of this change. The task and mission of agriculture has been changing over the last few decades. No longer are the productivity and economic aims prevailing, but environmental targets have been adopted and a good combination of these differing goals is needed. Agriculture is no longer seen as just a food production-oriented activity; it's also an environmental activity. And it is becoming more accepted that it is fulfilling a social aim. Agricultural sciences have seen a similar evolution. In the past they were largely one-dimensional, looking at only one aspect. Now we have the potential for taking a broader view as the science-based understanding and insight enables us to manipulate these agricultural systems for different objectives. We have seen how some of the old concepts can be replaced by new perspectives.

My second point is that we are moving away from traditional agronomy to production ecology. We are moving away from marginal changes with fine-tuning into structural changes in land use and production techniques. We're moving away from looking solely at individual fields to considering regional levels. We are moving away from a purely reductionistic science to an integrated and synthetic science. All these new developments are possible due to new technologies and insights. Precision agriculture can play an important role in this, and has the possibility to develop further.

My third point is that precision agriculture is probably more a direction of research and activity than a destination.

We've discussed clearly that several tools are now available but these should be improved and fine-tuned to the specific objectives we have in mind. At the field level, in the past we had the 'green fingers' of the farmers, now we have the potential for developing the 'green brain' of the farmer through science. This is done by designing the appropriate sampling procedures and methods which are rapid, accurate and take into account changes in time and space. We have made clear that the second element which is important is the characterization of heterogeneity in space. We have to know the spatial patterns and how they change, and this is an area where geostatistics can be of use. We have seen that modelling is not usable in all cases. We have to consider models with care. Very often explanatory models aren't much use for decision making at the field level. Regression-type models may be useful for prediction, but summary-type models may be of use to direct operational decision making. Basic information on the functioning of crops and soils is important to facilitate good science-based activity in precision agriculture.

The fourth point I would like to stress is that it is not enough to look just at the field level: it is also important to look at the level of farming systems. Here we have to take into account that we are studying various farming systems, including 'hi-tech' agriculture in Europe and the USA, and also the more elementary types of agriculture seen in many parts of Africa. It is also necessary to consider the socio-economic context within which these farming systems operate.

At the regional level, topographic and soil and climate data are readily available and can be helpful in exploratory studies of land use at different locations in order to meet the objectives we formulated earlier. Here we have another recipient of our knowledge: the policy makers.

The fifth point I'd like to mention is that precision agriculture in this broader sense has something to offer, but if we want to stay in business it must fulfil five minimum requirements. First, there must be a continuous updating of the information we are delivering to the users. Secondly, this information must be upgraded: there should be on-going improvement of the quality of the advice and information supplied. Thirdly, precision agriculture should be broadened in the sense that the information that is made available is biologically sound and continuously improved. We must consider not just the yield-limiting factors such as water and nutrients, but also the crop-reducing factors such as pests and diseases. To make this operational requires further interaction. The fourth requirement is that precision agriculture must be cost-effective. Finally, it should be broadened to include not only operational decisions but also work within the framework of tactical and strategic decision making.

There are also some dangers in the development of precision agriculture. It will fail if it is driven too much by technology, methodology or science, or it relies too heavily on inappropriate or insufficiently validated models, or if it is too package-oriented. The danger in scientific communities is that we are often too much science or technology driven: here we must be problem-oriented and driven by demand.

If we can meet these requirements and steer clear of the dangers, I see a bright future for precision agriculture.

Index of contributors

Non-participating co-authors are indicated by asterisks. Entries in bold type indicate papers; other entries refer to discussion contributions.

Indexes compiled by Liza Weinkove

Subject index